T0296444

The ability to grow cells in culture is an important and recognised part of biomedical research, but getting the culturing conditions correct for any particular type of cell is not always easy. These books aim to overcome this problem. The conditions necessary to culture different types of cell are clearly and simply explained in different volumes. Each volume covers a particular type of cell, and contains chapters by recognised experts explaining how to culture different lineages of the cell type. There is also a volume on general techniques in cell culturing. These practical handbooks are clearly essential reading for anyone who uses cell culture in the course of their research.

Marrow stromal cell culture contains chapters on the marrow stromal cell system, the bone marrow stroma *in vivo*, bone marrow stromal precursor cells, human bone-derived cells, marrow stromal adipocytes, osteoblast lineage in experimental animals, chondrocytes, and osteogenic potential of vascular pericytes.

Jon Beresford is Senior Lecturer in the School of Pharmacy and Pharmacology of the University of Bath and Head of Bone Research at Bath Institute for Rheumatic Diseases. Maureen Owen is former Director of the MRC Bone Research Laboratory in the Nuffield Department of Orthopaedic Surgery at the University of Oxford.

Marrow stromal cell culture

Handbooks in Practical Animal Cell Biology

Series editor:
Dr Ann Harris
Institute of Molecular Medicine, University of Oxford

The ability to grow cells in culture is an important and recognised part of
biomedical research, but getting the culturing conditions correct for any
particular type of cell is not always easy. These books aim to overcome this
problem. The conditions necessary to culture different types of cell are
clearly and simply explained in seven different volumes. Each volume
covers a particular type of cell, and contains chapters by recognised experts
explaining how to culture different lineages of the cell type. There is also a
volume on general techniques in cell culturing. These practical handbooks
are clearly essential reading for anyone who uses cell culture in the course
of their research.

Already published in the series:

Epithelial cell culture, edited by A. Harris

Endothelial cell culture, edited by R. Bicknell

General techniques of cell culture, by M. Harrison and I. Rae

Forthcoming in the series:

Haemopoietic and lymphoid cell culture,
edited by M. Dallman and J. Lamb

Endocrine cell culture, edited by S. Bidey

Marrow stromal cell culture

Edited by

Jon N. Beresford

and

Maureen E. Owen

CAMBRIDGE
UNIVERSITY PRESS

CAMBRIDGE UNIVERSITY PRESS
Cambridge, New York, Melbourne, Madrid, Cape Town, Singapore, São Paulo, Delhi

Cambridge University Press
The Edinburgh Building, Cambridge CB2 8RU, UK

Published in the United States of America by Cambridge University Press, New York

www.cambridge.org
Information on this title: www.cambridge.org/9780521589789

© Cambridge University Press 1998

First published 1998

A catalogue record for this publication is available from the British Library

Library of Congress Cataloguing in Publication data

Marrow stromal cell culture / edited by Jon N. Beresford and Maureen
E. Owen
    p.   cm. – (Handbooks in practical animal cell biology)
Includes bibliographical references and index.
ISBN 0 521 58021 8 (hardback). – ISBN 0 521 58978 9 (paperback)
1. Bone marrow – Cultures and culture media – Handbooks, manuals,
etc.   2. Bone marrow cells – Handbooks, manuals, etc.   I. Beresford,
Jon N.   II. Owen, Maureen E.   III. Series.
QP88.2.M177   1998
611′.0184–dc21   97–30196   CIP

ISBN 978-0-521-58021-2 hardback
ISBN 978-0-521-58978-9 paperback

Transferred to digital printing 2009

Cambridge University Press has no responsibility for the persistence or
accuracy of URLs for external or third-party Internet websites referred to in
this publication, and does not guarantee that any content on such websites is,
or will remain, accurate or appropriate. Information regarding prices, travel
timetables and other factual information given in this work are correct at
the time of first printing but Cambridge University Press does not guarantee
the accuracy of such information thereafter.

Every effort has been made in preparing this book to provide accurate and up-to-date information
which is in accord with accepted standards and practice at the time of publication. Although case
histories are drawn from actual cases, every effort has been made to disguise the identities of the
individuals involved. Nevertheless, the authors, editors and publishers can make no warranties that
the information contained herein is totally free from error, not least because clinical standards are
constantly changing through research and regulation. The authors, editors and publishers therefore
disclaim all liability for direct or consequential damages resulting from the use of material
contained in this book. Readers are strongly advised to pay careful attention to information
provided by the manufacturer of any drugs or equipment that they plan to use.

Contents

Contributors

Jane E. Aubin
Department of Anatomy and Cell Biology, University of Toronto, Room 6255, Medical Sciences Building, 1 King's College Circle, Toronto, Ontario M5S 1A8, Canada

Jon N. Beresford
School of Pharmacy and Pharmacology, University of Bath, Claverton Down, Bath BA2 7AY, UK

Paolo Bianco
Lab Fisiopatologia dello Scheletro Anatomia Patologica, Department of Experimental Medicine and Pathology, La Sapienza University, Viale Regine Elena 324 (Policlinico), 00161 Rome, Italy

Fiorella Descalzi Cancedda
Centro di Biotecnologie Avanzate, Istituto Nazionale per la Ricerca sul Cancro, Largo Rosanna Benzi 10, 16132 Genova, Italy and Istituto Internazionale di Genetica e Biofisica, Consiglio Nazionale delle Ricerche, via Marconi 10, 80100 Napoli, Italy

Ranieri Cancedda
Centro di Biotecnologie Avanzate, Istituto Nazionale per la Ricerca sul Cancro, Largo Rosanna Benzi 10, 16132 Genova, Italy and Dipartimento di Oncologia Clinica e Sperimentale, Universita di Genova, Largo Rosanna Benzi 10, 16132 Genova, Italy

Ann E. Canfield
Wellcome Trust Centre for Cell-Matrix Research, Department of Medicine, 2.205 Stopford Building, University of Manchester, Manchester M13 9PT, UK

Beatrice Dozin
Centro di Biotecnologie Avanzate, Istituto Nazionale per la Ricerca sul Cancro, Largo Rosanna Benzi 10, 16132 Genova, Italy

Jeffrey M. Gimble

Department of Surgery, University of Oklahoma Health Science Center, PO Box 26901, 920 Stanton L Young Blvd WP2140, Oklahoma City OK 73190, USA

Stephen E. Graves

Department of Orthopaedics and Trauma, University of Adelaide, Royal Adelaide Hospital, North Terrace, Adelaide 5000, Australia

Stan Gronthos

Matthew Roberts Laboratory, Leukaemia Research Unit, Division of Haematology, Hanson Centre for Cancer Research, IMVS, PO Box 14, Rundle Mall, Adelaide 5000, Australia

Roger Gundle

Nuffield Department of Orthopaedic Surgery, Nuffield Orthopaedic Centre, Headington, Oxford OX3 7LD, UK

Alexa Herbertson

Faculty of Dentistry, University of Toronto, Toronto, Ontario M5S 1A8, Canada

Maureen E. Owen

Medical Research Council Bone Research Laboratory, Nuffield Orthopaedic Centre, Headington, Oxford OX3 7LD, UK

Mara Riminucci

Deptartment of Experimental Medicine, University of L'Aquila, Italy

Ana M. Schor

Cell and Molecular Biology Unit, Dental School, University of Dundee, Dundee DD1 4HR, UK

Joanne Screen

School of Pharmacy and Pharmacology, University of Bath and Bath Institute for Rheumatic Diseases, Trim Bridge, Bath BA1 1HD, UK

Paul J. Simmons

Matthew Roberts Laboratory, Leukaemia Research Unit, Division of Haematology, Hanson Centre for Cancer Research, IMVS, PO Box 14, Rundle Mall, Adelaide 5000, Australia

Karina Stewart

School of Pharmacy and Pharmacology, University of Bath and Bath Institute for Rheumatic Diseases, Trim Bridge, Bath BA1 1HD, UK

Preface to the series

The series Handbooks in Practical Animal Cell Biology was born out of a wish to provide scientists turning to cell biology, to answer specific biological questions, the same scope as those turning to molecular biology as a tool. Look on any molecular cell biology laboratory's bookshelf and you will find one or more multivolume works that provide excellent recipe books for molecular techniques. Practical cell biology normally has a much lower profile, usually with a few more theoretical books on the cell types being studied in that laboratory.

The aim of this series, then, is to provide a multivolume, recipe-book-style approach to cell biology. Individuals may wish to acquire one or more volumes for personal use. Laboratories are likely to find the whole series a valuable addition to the 'in house' technique base.

There is no doubt that a competent molecular cell biologist will need 'green fingers' and patience to succeed in the culture of many primary cell types. However, with our increasing knowledge of the molecular explanation for many complex biological processes, the need to study differentiated cell lineages *in vitro* becomes ever more fundamental to many research programmes. Many of the more tedious elements in cell biology have become less onerous due to the commercial availability of most reagents. Further, the element of 'witchcraft' involved in success in culturing particular primary cells has diminished as more individuals are successful. The chapters in each volume of the series are written by experts in the culture of each cell type. The specific aim of the series is to share that technical expertise with others. We, the editors and authors, wish you every success in achieving this.

ANN HARRIS
Oxford, July 1995, March 1997, October 1997

Preface

Culture of marrow stromal cells cannot be covered comprehensively in a single small book. Our selection of subject matter for the different chapters was biased towards the cell lines of the marrow stromal system, which are more directly concerned with osteogenic tissues. Because of space, all authors were limited in the number of references allowed. Often a relevant review or recent paper which provides access to the previous literature has been cited as a best compromise. We apologise to those whose papers have not been quoted directly and we recognise and fully appreciate the importance of their work.

This book is dedicated to Alexander Friedenstein (1924–97), of the Gamaleya Institute, Moscow, who pioneered the subject of marrow stromal cell biology and made unrivalled contributions over many decades.

The Editors

1

The marrow stromal cell system

Maureen E. Owen

Introduction

The stromal tissue of the bone marrow consists of a fine mesh of blood vessels held in a network of cells and extracellular matrix (ECM); it forms a continuum with the soft connective tissue adjacent to endosteal and periosteal bone surfaces and within Haversian canals. Marrow stromal tissue provides physical support and nourishment for the haemopoietic stem and progenitor cells and this led to the concept of a haemopoietic microenvironment (HME) (Dexter, Allen & Lajtha, 1977). In addition, the osteogenic potential of marrow stromal tissue has been well known for many decades (for review see Burwell, 1994).

The term 'stromal cells' has been used for the partially defined population of cells which make up the adherent cell layer in *in vitro* long-term bone marrow cultures (LTBMC) (Dexter, 1982). In this chapter, the term stromal is restricted to non-haemopoietic cells of mesenchymal origin and does not include macrophages and endothelial cells, which are also components of the adherent layer. There are three main cellular systems in the bone marrow; haemopoietic, endothelial and stromal, which are histogenetically distinct with no common precursor in the post-natal animal. The hypothesis of a marrow stromal cell system was based on analogy with the haemopoietic system (Owen, 1985); in essence, it is proposed that there are marrow stromal stem cells (MSSC) present within the bone marrow, able to generate the cell lines giving rise to the fibrous–osteogenic tissues of the skeleton and the stromal tissues of the HME (Friedenstein, 1980,1990; Owen, 1988; Owen & Friedenstein, 1988).

The stromal or soft connective tissue of the bone marrow includes an outer capsule of bone lining cells consisting of osteoblasts and preosteoblasts. Endothelial cells line the inner surfaces of the vascular sinuses and on the

1

abluminal side there is a thick adventitial cell layer, which contains mainly reticular, adipocytic and smooth muscle cells. Reticular cells (also called reticular fibroblasts), which are the dominant cell type of marrow stroma, are broad in shape and have extensive cytoplasmic processes, which fuse forming a syncytium throughout the extravascular space; elongated fibroblasts are present in small numbers. Collagen is a major constituent of the ECM; types I and III are distributed throughout interstitial and perivascular tissues, and type IV, a constituent of basal laminae, is associated mainly with vessel walls (Bentley, Alabaster & Foidart, 1981).

Stem cells are defined as; able to self-renew, multipotential and capable of regenerating tissue after injury. The high regenerative capacity of marrow stromal cells was apparent from early studies demonstrating the generation of a bone marrow organ after transplantation of marrow to an ectopic site (Tavassoli & Crosby, 1968) and from tissue regeneration following marrow ablation (references in Suva et al., 1993). Heterotopic transplantation of marrow stromal cells, either in an open site or within a diffusion chamber (DC), is the basis for current in vivo assays of stromal stem and progenitor cells. In open transplant (OT) it is possible to assay for generation of all the fibroblastic, osteogenic and stromal lines of the HME. In the DC, tissues from the host do not enter the chamber and the system, as used at present, assays for generation of fibroblastic, bone and cartilage tissues (Friedenstein, Chailakhjan & Gerasimov, 1987; Owen, 1988).

In vivo assay of marrow stromal stem and progenitor cells

Marrow cells, prepared as a single cell suspension and cultured in vitro with fetal calf serum (FCS), form fibroblastic colonies each derived from a single cell or colony forming unit-fibroblastic (CFU-F). That CFU-F are non-cycling in vivo and the colonies clonally derived, was confirmed by a number of tests (Friedenstein, 1976, 1990; Falla et al., 1993). Assay in OT in vivo of the tissue formed from cloned populations derived from individual CFU-F demonstrated that MSSC (able to form all the stromal lines of a bone and marrow organ) and precursors with more restricted potential (able to form either a soft fibrous or osteogenic tissue only, here designated CFU-fib and CFU-O, respectively), were present among the CFU-F (Friedenstein,1980). With the DC assay, it was shown that bone and cartilage have a common precursor in marrow, and also that fibroblasts cultured from CFU-F have an extensive capacity for self-renewal and for osteogenesis (Friedenstein et al., 1987). It was calculated that cells derived from the marrow of a fragment of rabbit pelvic bone, 0.5 g weight, are capable of giving rise to 30 kg of bone

(Friedenstein, 1990). The possible clinical applications of this cell population for tissue repair are a subject of active investigation (Connolly, 1995).

In vitro assay of marrow stromal stem and progenitor cells

In the standard conditions used for assay of CFU-F in vitro, FCS is the only serum supplement; the cells have an elongated, fibroblastic morphology, are commonly called marrow fibroblasts, and synthesise collagen types I and III but not type IV (Friedenstein, 1976, 1990; Castro-Malaspina et al., 1980; Lim et al., 1986; Singer et al., 1993). The CFU-F assay provides a quantitative measure of a compartment of relatively undifferentiated stromal precursors, which can be considered as a developmentally structured continuum of cells with a declining capacity for self-renewal. Colonies formed in the assay are heterogeneous in size, morphology and potential for differentiation (Owen, 1988) and, for this and other reasons, the relevance of CFU-F numbers may be difficult to interpret.

In the haemopoietic system, in vitro clonogenic assays, able to detect progenitor cells by their ability to give rise to colonies of morphologically recognisable differentiated progeny, were instrumental in advancing the understanding of cell differentiation in that system. This approach has been less feasible for the marrow stromal system, due mainly to the lack of specific colony stimulating factors (CSFs) and markers for the different cell lines. The difficulties are indicated by the fact that the best method of identifying the majority of stromal cell lines at present, is by the mixture of collagen types and other ECM molecules synthesised, which is often neither specific nor known in detail. Furthermore, most of the wide range of hormones, cytokines and growth factors, which act on marrow stromal lines, are also common to stromal, haemopoietic and many cell types in other organs. Nevertheless, a number of marrow stromal progenitor cells have been identified using different conditions of culture.

An in vitro colony assay was developed for osteogenic progenitors (details and references in Aubin & Herbertson, Chapter 6). When marrow stromal cells are grown in vitro in osteogenic inducing conditions (OGC), which include FCS, an appropriate level of glucocorticoid, a source of phosphate and vitamin C, discrete three-dimensional mineralising nodules of bone tissue develop within a fibrous tissue. A single cell, or colony forming unit-osteogenic (CFU-O), is sufficient for production of one bone nodule and CFU-O have a limited capacity for self-renewal, typical of progenitors with restricted potential.

Putative clonal progenitors for cell lines of the HME have been identified when marrow cells are cultured in LTBMC conditions; these usually include glucocorticoid and sera other than FCS, e.g. horse or homologous serum. In a study of human marrow cells using LTBMC conditions, colonies with reticular–fibroblastic morphology were formed, many with lipid inclusions, each from a single cell or colony forming unit reticular fibroblastic (CFU–RF) (Lim et al., 1986). In contrast to CFU-F, CFU-RF produced laminin (LN) and collagen type IV in addition to types I and III; both were negative for factor VIII. Reticular fibroblasts are thought to synthesise reticulin, the sparse collagenous matrix in marrow, and to be precursors of the adipocytic lineage (Bianco & Riminucci, Chapter 2). It is proposed that CFU-RF may be a clonal progenitor for the reticular–fibroblastic–adipocytic tissue which is found throughout marrow and is a major part of the HME.

The identification of a monoclonal antibody STRO-1, which binds specifically to all CFU-F in human marrow, was a significant advance (Gronthos, Graves & Simmons, Chapter 3). STRO-1 positive cells isolated by FACS have been cultured in different conditions in vitro; in FCS a non-mineralised soft connective tissue is formed, in OGC conditions a mineralised fibrous-osteogenic tissue is produced, sometimes with adipogenesis, and in LTBMC conditions a number of differentiated cells, fibroblastic cells, pre-adipocytes and smooth muscle cells, all of which expressed the STRO-1 antigen, are generated. The data support the conclusion that the CFU-F population contains stem and progenitor cells for fibrous, osteogenic and other stromal cell lines. However, note (Gronthos et al., Chapter 3), that there may be stromal precursor cells in vivo which are not identified in current in vitro assays.

Other cells associated mainly with the microvasculature in stromal tissues, are pericytes and smooth muscle(sm) cells (Schor & Canfield, Chapter 8). These two cell types synthesise type IV collagen and LN and share many phenotypic markers. Although progenitors (CFU-sm) have not been isolated nor relationships to the other stromal cell lines elucidated there is evidence that these cells contribute to the HME (Galmiche et al., 1993). Pericytes have rarely been reported in bone marrow, and the possibility that in marrow they are replaced by specialised perisinusoidal reticular cells has been suggested (Nehls & Drenckhahn, 1993).

Plasticity of marrow stromal cells

There is growing evidence from in vitro studies for a high degree of plasticity among stromal cell lines. Evidence for a common progenitor for the three tissues, adipocytic, fibroblastic and osteogenic, and a direct demonstration of the ability of stromal cells to switch from one phenotype to another was first obtained in studies using rodent marrow (discussion and references to Bennett *et al.*, 1991 and Beresford *et al.*, 1992 in Chapters 2, 4 and 5). These studies demonstrated that highly differentiated adipocytes are able to dedifferentiate *in vitro* back to a less differentiated, more proliferative fibroblastic precursor, and then to express an osteogenic phenotype when cultured in a DC *in vivo*. Other examples of plasticity are seen in chondrocytes and pericytes; cells which may also dedifferentiate and express an osteogenic phenotype under certain conditions *in vitro* and in a DC *in vivo* (Cancedda, Cancedda & Doizin, Chapter 7, Schor & Canfield, Chapter 8). These results raise the question of commitment. Clearly, *in vitro*, commitment is reversible; whether *in vivo*, there is irreversible commitment in this system under normal physiological conditions, and at what stage, is not known. In pathological conditions, however, it seems evident that cells can switch from one phenotype to another depending upon the local microenvironment, probably via an earlier proliferative stage (Bianco & Riminucci, Chapter 2).

Discussion

It is unlikely that MSSC with developmental potential have a major role in normal bone and marrow physiology in the post-natal animal and it is reasonable to postulate that they are an emergency reserve for crisis situations. Maintenance of cell numbers on a daily basis is probably regulated at the level of less primitive fibroblastic precursors. The case for the existence of marrow stromal stem cells rests largely upon the demonstration, that the cell progeny from a single stromal cell is able to develop into a bone marrow organ containing the full spectrum of stromal cell types in the *in vivo* OT assay. A critical issue is the clonality of the cell progeny, and stringent proof of this has yet to be obtained, e.g. by the application of techniques able to mark unequivocally single stromal stem cells and to follow their progeny, such as have been applied in other systems. Nevertheless, the evidence for a marrow stromal cell system and marrow stromal stem cells, although circumstantial, is very strong.

Our knowledge of the multipotentiality of stem cells and of the hierarchy of cell lineages in the stromal system *in vivo*, is only as good as the method of assay. In the OT assay *in vivo* (Friedenstein, 1980, 1990) tissue formation mimics the sequential stages in embryonic development (Bianco & Riminiccu, Chapter 2) and one of the best descriptions of this is to be found in Tavassoli and Crosby, 1968. The proposed hierarchy for tissue

	(1)	(2)	(3)
(a) MSSC	Fibrous tissue	Fibrous-osteogenic tissue	Fibrous-osteogenic and stromal tissues of the HME Vessel ingrowth
(b)	+ I, + III + VN, + FN - IV, - LN - smA - FVIII	+ I, + III	+ I, + III + IV, + LN + smA + FVIII
			CFU-sm CFU-RF
	CFU-fib	CFU-O CFU-fib	CFU-O CFU-fib
(c) MSSC	MSSC	MSSC	MSSC

Fig. 1.1. Tissue and cell lineage hierarchy in the marrow stromal system, based on development of a bone marrow organ after transplant of a small marrow fragment or a cloned population from a single stromal stem cell, in the *in vivo* OT assay. (a) From left to right, stages (1) to (3) of tissue generated by MSSC (see text). (b) Collagen types, ECM components and other markers at different stages. + present − not present. (c) Stem cells and lineage progenitors (CFUs) present at different stages. CFU-fib and CFU-O, progenitors of fibrous-osteogenic tissues appear earlier in the hierarchy than CFU-RF and CFU-sm, putative progenitors of the HME. MSSC and CFUs are heterogeneous cell populations.

differentiation and lineage progenitors (CFUs) in the marrow stromal system, based on these observations and supported by a range of *in vivo* and *in vitro* data, is shown in Fig. 1.1.

In stage (1) a fibrous tissue is formed which contains both collagen types I and III, is positive for vimentin (VN) and fibronectin (FN) and negative for collagen type IV, laminin (LN) smooth muscle actin (smA) and factor VIII. Type III predominates during the early phase of fibroblastic proliferation, and then as osteogenic differentiation takes place, stage (2), type I increases in conjunction with the formation of a lattice of trabecular bone (Lane *et al.*, 1986; Zhou *et al.*, 1995). By stage (3) there is a fully formed bone and marrow organ with a functioning HME, the cells of which synthesise collagen type IV, laminin (LN) and smA in addition to collagens I and III. Endothelial cells of ingrowing vessels are positive for factor VIII. Previous work has shown that, both the cells of the fibrous–osteogenic tissue *and* those of the stromal lines of the HME originate from the donor transplant, whereas the haemopoietic and endothelial cells are from the host (for review see Owen, 1988; Friedenstein, 1990; Schor & Canfield, Chapter 8). It follows therefore that stromal lines of the HME are generated from a cell population which earlier had given rise to fibrous-osteogenic tissues (Fig. 1.1).

Cell terminology in this subject has not been standardised. 'Stromal cells' has long been used for precursors of the HME (Singer, Keating & Wight, 1985) and the terms fibroblast, reticular fibroblast and reticular cell, are often used interchangeably. Two categories of fibroblastic cell in marrow are distinguishable simply by collagen types synthesised, as well as in many other respects. First, marrow fibroblasts, which synthesise collagen types I, III and V but *not* IV and are the precursors of fibrous-osteogenic tissue. Secondly, marrow reticular fibroblasts (also called reticular cells), which synthesise collagen type IV in addition to types I, III and V (Lim *et al.*, 1986; Friedenstein, 1990; Singer *et al.*, 1993), and are the supposed precursors of some or all of the stromal lines of the HME. It follows that marrow reticular cells are the progeny of marrow fibroblasts and that progenitors for fibrous-osteogenic tissue occur earlier in the hierarchy than those for the HME (Fig. 1.1). Consistent with this idea is the demonstration that murine marrow fibroblasts positive for collagen types I and III and negative for type IV, support hemopoiesis when cultured under LTBMC conditions (Brockbank *et al.*, 1986). Furthermore, among cell lines established from murine marrow stromal cells, those which synthesise collagen type IV and LN tend to support haemopoiesis *in vitro*, whereas those which synthesise only I and III do not (Kirby & Bentley, 1987). The commitment of MSSC and /or their progeny to differentiate into cell lines of the HME takes place within a very complex environment, where bone formation, vessel invasion, the arrival of haemopoietic cells, resorption and remodelling of bone and haemopoiesis are taking place. The order in which the different processes occur is uncertain and time-wise they are likely to overlap; vascular invasion may be a key event (Fig. 1.1).

Future progress requires more markers for lineage identification and development of new assay systems and models for studying differentiation, both *in vitro* and *in vivo**. The potential of the DC system for identifying the tissues differentiating from stem and progenitor cells, and the use of the punctured DC which could give information on conditions which induce the HME, have been little explored (Shimomura, Yoneda & Suzuki, 1975; Bab & Einhorn, 1993). A long-term objective is the transplantation of marrow stromal cells for gene therapy; on present knowledge, conditioning

8 • Maureen E. Owen

of the site and prior treatment of the cells to be transplanted may be important for successful homing and graft. *In vitro* methods for expansion of
stromal stem cells with a view to their use in bone healing is an area on which
much current work is being concentrated.

*Note added in proof

A new antibody (HOP-26) reacting with human marrow stromal cells at
early stages of differentiation has been described recently.

References

Bab, I.A. & Einhorn, T.A. (1993). Regulatory role of osteogenic growth polypep-
tides in bone formation and haemopoiesis. *Crit. Rev. Eukaryotic Gene Expression*,
3(1), 31–46.
Bentley, S.A., Alabaster, O. & Foidart, J.M. (1981). Collagen heterogeneity in
normal human bone marrow. *Br. J. Haemat.*, 48, 287–91.
Brockbank, K.G.M., de Jong, J.P., Piersma, A.H. & Voerman, J.S.A. (1986).
Haemopoiesis on purified bone-marrow derived reticular fibroblasts *in vitro*.
Exp. Hemat., 14, 386–94.
Burwell, R.G. (1994). The importance of bone marrow in bone grafting. In *Bone
Grafts, Derivatives and Substitutes*, ed. M.R. Urist, B.T. O'Connor & R.G.
Burwell. Butterworth–Heinemann Ltd.
Castro-Malaspina, H., Gay, R.E., Resnick, G., Kapoor, N., Chiariere, D.,
McKenzie, S., Broxmeyer, H.E. & Moore, M.A. (1980). Characterization of
human bone marrow fibroblast colony forming cells (CFU-F) and their
progeny. *Blood*, 56, 289–301.
Connolly, J.F. (1995). Injectable bone marrow preparations to stimulate osteogenic
repair. *Clin. Orthop.*, 313, 8–18.
Dexter, T.M. (1982). Stromal cell associated haemopoiesis. *J. Cell Physiol. Suppl.*, 1,
87–94.
Dexter, T.M., Allen, T.D. & Lajtha, L.G. (1977). Conditions controlling the proliferation of haemopoietic stem cells *in vitro*. *J. Cell Physiol.*, 91, 335–44.
Falla, N., Van Vlasselaer, P., Bierkens, J., Borremans, B., Schoeters, G. & Van Gorp,
U. (1993). Characterization of a 5-fluorouracil-enriched osteoprogenitor
population of the murine bone marrow. *Blood*, 82, 3580–91.
Friedenstein, A.J. (1976). Precursor cells of mechanocytes. *Int. Rev. Cytol.*, 47,
327–55.
Friedenstein, A.J. (1980). Stromal mechanisms of bone marrow: cloning *in vitro* and
retransplantation *in vivo*. In *Immunology of Bone Marrow Transplantation* ed. S.
Thienfelder, H. Rodt & H.J. Kolb, pp. 19–29. Berlin: Springer-Verlag.

(1990). Osteogenic stem cells in the bone marrow. In *Bone and Mineral Research*, vol. 7, ed. J. Heersche & J.A. Kanis, pp. 243–272. Elsevier.

Friedenstein, A. J., Chailkhjan, R.K. & Gerasimov, U.V. (1987). Bone marrow osteogenic stem cells: *in vitro* cultivation and transplantation in diffusion chambers. *Cell Tissue Kinet.*, **20**, 263–72.

Galmiche, M.C., Koteliansky, V.E., Briere, J., Herve, P. & Charbord, P. (1993). Stromal cells from human long-term marrow cultures are mesenchymal cells that differentiate following a vascular smooth muscle pathway. *Blood*, **82**, 66–76.

Joyner, C.J., Bennett, A. & Triffitt, J.T. (1997). *Bone*, **21**, 1–6.

Kirby, S.L. & Bentley, S.A. (1987). Proteoglycan synthesis in two murine bone marrow stromal cell lines. *Blood*, **70**, 1777–83.

Lane, J.M., Suda, M., von der Mark, K. & Timpl, R. (1986). Immunofluorescent localization of structural collagen types in endochondral fracture repair. *J. Orthop. Res.*, **4**, 318–29.

Lim, B., Izaguirre, C.A., Aye, M.T., Huebsch, L., Drouin, J., Richardson, C., Minden, M.D. & Messner, H. A. (1986). Characterization of reticulofibroblastoid colonies (CFU-RF) derived from bone marrow and long-term marrow culture monolayers. *J. Cell Physiol.*, **127**, 45–54.

Nehls, V. & Drenckhahn, D. (1993). The versatility of microvascular pericytes: from mesenchyme to smooth muscle. *Histochemistry*, **99**, 1–12.

Owen, M. (1985). Lineage of osteogenic cells and their relationship to the stromal system. In *Bone and Mineral Research*, ed. W.A. Peck, vol 3, pp. 1–25. Elsevier. (1988). Marrow stromal stem cells. *J. Cell Sci. Suppl.*, **10**, 63–76

Owen, M. & Friedenstein, A.J. (1988). Stromal stem cells: marrow derived osteogenic precursors. *Ciba Foundat. Symp.*, **136**, 42–60.

Shimomura, Y., Yoneda, F., & Suzuki, F. (1975). Osteogenesis by chondrocytes from growth cartilage of rat rib. *Calc. Tiss. Res.*, **19**, 179–88.

Singer, J.W., Keating, A., & Wight, T.N. (1985). Human haematopoietic microenvironment. In *Recent Advances in Haematology*, ed. A.V. Hoffbrand, No. 4, pp. 1–25, Churchill Livingstone.

Singer, J.W., Slack, J.L., Lilly, M.B. & Andrews, D.F. (1993). Marrow stromal cells: response to cytokines and control of gene expression. In *The Hematopoietic Microenvironment*, ed. M.W. Long & M.S. Wicha, pp. 127–51. Baltimore and London: Johns Hopkins University Press.

Suva, L.J., Seedor, J.G., Endo, N., Quartuccio, H.A., Thomson, D.D., Bab, I. & Rodan, G.A. (1993). Pattern of gene expression following rat tibial marrow ablation. *J. Bone Min. Res.*, **8**, 379–88.

Tavassoli, M. & Crosby, W.H. (1968). Transplantation of marrow to extramedullary sites. *Science*, **161**, 54–6.

Zhou, H., Choong, P.F.M., Henderson, S., Chou, S.T., Aspenberg, P., Martin, T.J. & Ng, K.W. (1995). Marrow development and its relationship to bone formation *in vivo*: a histological study using an implantable titanium device in rabbits. *Bone*, **17**, 407–15.

2

The bone marrow stroma *in vivo*: ontogeny, structure, cellular composition and changes in disease

Paolo Bianco and Mara Riminucci

Introduction

The word stroma is generally used in anatomy and histology to signify the supporting connective tissue associated with the dominant functional tissue (the parenchyma) in an organ. There is perhaps no other organ, however, for which the original meaning (στρομα = mattress, that which one rests or lies upon) of the word stroma applies as appropriately as for the stroma of the bone marrow. Here, maturing precursors of blood cells rest directly upon surfaces provided by the 'stromal' cells. This is a notable peculiarity of the bone marrow stroma: it is largely made of cells and cell surfaces, rather than of physically conspicuous extracellular matrix components, such as the collagenous scaffolds holding parenchymal tissues together in most other organs. This reflects the special nature and function of the bone marrow stroma with respect to haematopoiesis, i.e. not just a system of physical support, but the repository of a host of cell-derived cues and signals driving the commitment, differentiation and maturation of haematopoietic cells.

Different definitions of the bone marrow stroma result in different concepts of its identity and cellular composition, and in some confusion. Anatomically, the stroma of the mammalian post-natal bone marrow is the three-dimensional network of cell surfaces holding maturing blood cells together in the extravascular space. Four main cell types comprise this network: macrophages, adipocytes, osteogenic cells near bone surfaces, and cells commonly referred to as 'reticular' cells. In long-term bone marrow cultures, stromal cells are the heterogeneous adherent cell layer providing the conditions for growth and differentiation of haematopoietic cells *in vitro*. This cultured stroma includes cells of fibroblastic morphology, macrophages, lipid-containing cells (adipocytes), endothelial cells and smooth muscle cells.

The role of individual cell types in providing an *in vitro* microenvironment for haematopoiesis, has not been fully elucidated, and some of the cell types which grow in such systems *in vitro* may not be functional components of the stroma with regard to haematopoiesis. Smooth muscle cells and endothelial cells are located within vessel walls and are not found in the extravascular space of the bone marrow where haematopoiesis occurs in mammals, and are not part of the stroma in a strictly anatomical sense. Evidence from a number of experimental approaches, human and animal pathology, and transplantation studies have clearly shown that the 'fibroblastic' stromal cells are ontogenetically unrelated to haematopoietic cells. Although the major cells of the marrow stroma are often regarded as 'fibroblastic' in nature, they differ, both functionally and phenotypically, from other fibroblasts. The haematopoietic microenvironment in the normal bone marrow contains sparse amounts of collagen. Deposition of conspicuous amounts of collagen fibres (fibrosis) in the marrow is always a response to injury. True fibroblasts (here taken as connective tissue cells involved in the deposition of significant amounts of extracellular collagenous matrix) are not regular residents of the bone marrow stroma, but appear to be present in certain pathological states.

In this chapter, bone marrow stromal cells are defined on the basis of the following criteria: (a) they are found in the extravascular compartment of the mammalian bone marrow, (b) they participate in providing physical and functional support for haematopoietic cells, (c) they are not of haematopoietic lineage, and (d) they are members of the marrow stromal system (Owen 1985). The list of 'cell types' resulting from this definition obviously reflects our ability to recognise them in samples of intact bone marrow. Technical difficulties related to the proper handling of marrow tissue for histology at the appropriate level of sophistication, and the elusiveness of the morphology of marrow stromal cells *in vivo* have certainly hampered progress in this area. An *in vivo* perspective is, however, a desirable complement to *in vitro* studies, although it must be remembered that many of the properties displayed by cells in culture may not reflect the *in vivo* situation. For example, smooth muscle-specific actin isoforms are expressed by some marrow 'fibroblasts' in culture, yet only smooth muscle cells in arteriolar walls, seem to express this marker in the normal human and rat marrow *in vivo*. Caution is also in order when molecules whose expression is affected by *in vitro* culturing *per se*, such as integrins or other adhesion molecules, are considered.

The following pages are meant to provide a basic account of the histology and cell composition of the mammalian bone marrow stroma, rather than a comprehensive survey of phenotypic properties of marrow stromal cells. The latter would involve extensive reference to work on *in vitro* systems, whereas the scope of this chapter is rather to provide a view of stromal cells in their native environment. Emphasis will thus be placed on those structural, developmental, or pathological aspects of marrow biology that we regard as relevant to the understanding of cell differentiation within the marrow stromal system, and to the physiology and diseases of the skeleton.

The ontogeny of the bone marrow stroma

During development there is no bone marrow and no bone marrow stroma, before there is bone. The formation of marrow cavities in long bones of the limbs, which is sometimes taken as the watershed of embryonic and fetal life (Streeter, 1949), only occurs after a distinct layer of bone has formed around cartilage rudiments (Fig. 2.1(a)). This layer forms a continuous sheath around the mid-diaphyses of the bone rudiment (the bony collar), and is deposited by mature osteoblasts that differentiate from osteoprogenitor (osteogenic precursor) cells in the primitive periosteum (perichondrium). Osteogenic precursor cells and fully mature osteoblasts, competent to deposit a mineralising matrix, have already appeared when the bone marrow begins to form. The formation of marrow cavities depends on erosion of the bony collar and the underlying hypertrophic cartilage by osteoclasts (Fig. 2.1(b). Before haematopoiesis can settle in these forming cavities, a stromal scaffold is established as the first event in the formation of marrow. This occurs in association with the development of a system of dilated, varicose, thin-walled and basement membrane-free blood vessels (the primitive sinusoids), which originate from the branching of arterial vessels around and inside the bone rudiment. The primitive marrow stroma, not yet populated by haematopoietic cells, appears simply as a meshwork of branched cells. These cells occupy the spaces separating the outer walls of developing sinusoids from each other and from the forming endosteal bone surfaces (Fig. 2.1(b)), where they are located next to the layer of mature, bone-forming osteoblasts, a position equivalent to that occupied by committed osteoprogenitor cells in the primitive periosteum (Fig. 2.1(a)). The primitive marrow stroma and the layer of osteoprogenitor cells in the primitive periosteum are physically continuous with one another (Fig. 2.1(b)). This reflects the continuity of the vascular branches which come from the periosteum, to the endosteal surfaces of the forming shaft and to the diaphyseal marrow cavity. In a strict anatomical sense, the cellular coat of the vessels which grow into the forming marrow cavity, originates as an invagination of the osteoprogenitor cell layer in the primitive periosteum (Fig. 2.1). Vascular loops in the marrow cavity are thus coated with osteogenic cells, as are the internal endosteal surfaces of the bony walls of the cavity. The two linings are continuous with one another and with the primitive periosteum.

Primitive marrow stromal cells and osteoprogenitor cells are phenotypically similar (Bianco *et al.*, 1993*b*). Both are positive for alkaline-phosphatase (ALP) activity and negative for certain markers of mature osteogenic cells, such as bone sialoprotein (BSP). The histologically identifiable primitive stromal cells of the pre-haematopoietic marrow can reasonably be seen as the committed (ALP+) osteogenic precursor cells associated with the formation of endosteal bone surfaces. A pre-haematopoietic marrow also exists beyond fetal life. It is found in growing bone immediately underneath the epiphyseal growth plate, in the region of the primary spongiosa. Here, sinusoids develop between the endochondrally formed trabeculae, and primitive ALP+stromal cells are found between the osteoblasts and the sinus walls. The cells found to bind radiolabelled parathyroid hormone (PTH) in the primary spongiosa of growing rats (Rouleau, Mitchell & Goltzmann, 1988) are in fact ALP+, primitive stromal cells of the pre-haematopoietic marrow within the primary spongiosa. Like osteoprogenitor cells in the periosteum, primitive stromal cells of pre-haematopoietic marrow incorporate the thymidine analogue bromodeoxyuridine (BrdU) (Bianco *et al.*,1993*b*). In contrast, there appears to be little, if any, incorporation of BrdU in the stromal cells of haematopoietic marrow of growing animals, and none in the marrow associated with non-growing skeletal segments, in keeping with the known radioresistance of marrow stroma.

The development of a system of sinusoids provides a circulation with characteristics which permit seeding of the primitive marrow environment with blood-borne haematopoietic stem cells. This requires that blood flow slows significantly and that the vessel walls allow transendothelial migration of individual cells, the two specific characteristics that make sinusoids different from capillaries (Tavassoli & Yoffey, 1983). The development of sinusoids and the homing of haematopoietic stem cells (the seed) to the microenvironment of stromal cells of osteogenic potential (the soil), results in the formation of spaces that become filled with maturing haematopoietic cells, which otherwise would be fated to be filled with bone. It is suggested that the maturation of osteogenic precursor cells with ensuing bone formation may be under the negative control of haematopoietic cells, with the object of providing the necessary space for haematopoiesis. This can only be a matter of speculation at present, but, in principle, it is consistent with the negative effects on osteoblast differentiation exerted by certain haematopoietic cytokines (Gimble *et al.*, 1994; Gimble, Chapter 5).

Human disease provides examples of a mutual balance between osteo-

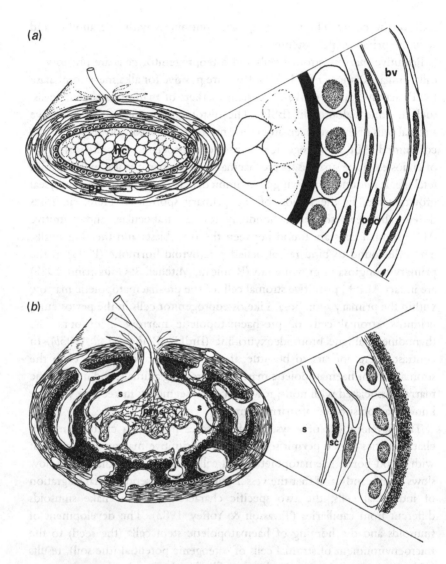

Fig. 2.1. Development of the bone marrow. (a) Diagram of a cross-section of the mid-diaphyseal region of a rib of a 16-day rat embryo. The cartilage rudiment has not yet been invaded by blood vessels, and the mid-diaphysis is occupied by hypertrophic cartilage (hc). This is surrounded by a collar of osteogenic cells, noted for the expression of ALP activity, which distinguishes them from the surrounding, ALP-negative, embryonic soft tissues. The osteogenic collar (primitive periosteum, pp) is permeated by thin-walled capillaries (blood vessels, bv), and is composed of an innermost layer of mature osteoblasts (o), and several outer layers of flattened osteogenic precursor cells (opc). These cells, while expressing ALP activity, do not express markers of mature osteoblasts, such as bone sialoprotein, but do

genesis and myelogenesis and supports the postulate there is competition for space between marrow and bone. The skeleton of children with congenital haemolytic anaemias undergoes remarkable changes in response to the abnormal expansion of the bone marrow (Ascenzi, 1976). This expansion takes place in order to compensate for the shortened lifespan of red blood cells, and in the marrow cavity of children this cannot occur at the expense of intramedullary spaces occupied by non-haemopoietic tissues, as happens in the case of an adult. As a result, the marrow gains space at the expense of the encasing bone tissues. In the skull, for example, cancellous or trabecular bone develops *in lieu* of compact bone, and in the vertebrae, native cancellous bone becomes, for the same reasons, severely porotic (Ascenzi, 1976).

The post-natal marrow stroma

The three-dimensional complexity of the post-natal bone marrow structure is not easily perceived in two-dimensional histological sections. Like developing marrow, post-natal marrow is the tissue filling the space between blood vessel walls and bone surfaces in the pores of cancellous bone (Fig. 2.2). The arteries branch within the marrow into a network of sinusoids that represent the terminal bed of a single vascular system shared by cancellous bone and marrow. As the arteries branch, so does the layer of connective tissue cells located on their outer surface (the adventitia). As they become arterioles, it is noted that the adventitia becomes enriched in ALP+cells. When the arteriole divides into sinusoids, these cells provide an adventitial layer covering the outer surface of the sinusoidal endothelium with no intervening base-

incorporate the thymidine analogue, BrdU. The mature osteoblasts, which do not incorporate BrdU, deposit the first layer of bone (the primitive bony collar, black in the enlarged view) *before* the invasion of the underlying cartilage begins.
(b) Diagram of a cross-section through the mid-diaphyseal region of the rib in an 18–19-day rat embryo. Formation of bone has progressed to the construction of primitive haversian arcades, encircling dilated thin-walled blood vessels reminiscent of sinusoids. Cartilage has been invaded, resorbed, and replaced by the primitive, pre-haematopoietic bone marrow. This consists of a network of sinusoid-like blood vessels (s), between which a meshwork of cells, the primitive marrow stroma (pms) has appeared. The system of sinusoids communicates extensively with vessels in the bony collar, and the primitive marrow stroma is physically continuous with the primitive periosteum. The primitive stromal cells (sc) like periosteal osteoprogenitor cells, express ALP activity and incorporate BrdU, and lie next to the osteoblasts (o) lining the forming endosteal surfaces. Haematopoiesis will develop once the haematopoietic stem cells reach the primitive stromal tissue via the sinusoidal circulation.

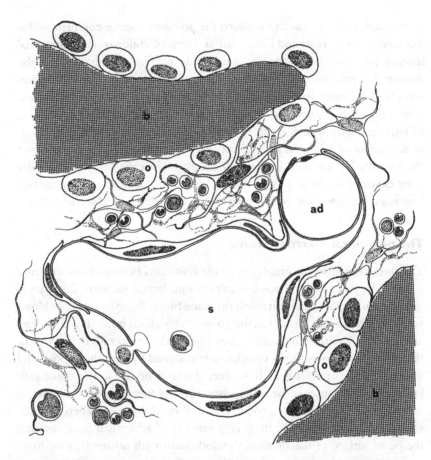

Fig. 2.2. Diagram of the structure of the post-natal marrow of cancellous bone. A network of sinusoids (s) represent the terminal branching of arteries of cancellous bone. The space between the sinusoidal walls and bone surfaces (extravascular compartment of the marrow, where haematopoiesis occurs in mammals) is occupied by maturing haematopoietic cells. These are held in a meshwork of cell surfaces provided by the stromal cells (sc), including those lining the sinusoidal outer wall, those lining bone (b), and those interspersed in the extravascular space. The diagram also shows the 'niches' provided by stromal cell processes to individual cells, and the trans-endothelial migration of cells across the sinus wall. This is the mechanism whereby mature blood cells leave the marrow, and circulating hematopoietic stem cells enter the marrow. ALP+ reticular cells are not phagocytic and are distinguishable from macrophages, which are other cells of stellate or 'reticular' morphology found among haematopoietic cells in the extravascular compartment. An adipocyte (ad) in adventitial position compresses the sinus wall.

ment membrane. This layer is composed of slender, elongated and attenuated cells known from earlier scanning electron microscopy studies as 'adventitial reticular cells' (Weiss, 1976; Weiss & Sakai, 1984). They are similar to cells found in the extravascular space and together they are called reticular cells. They extend long cell processes through the extravascular space, where they come in contact with, and hold, maturing blood cells, among which macrophages are also interspersed.

Because of their inconspicuous morphology, a complex branched shape with extensive slender cell processes, these reticular cells cannot be resolved in sections prepared for routine light microscopy. However, they can be imaged effectively by scanning electron microscopy (SEM) (Weiss & Sakai 1984), but cannot always be distinguished from macrophages, which also have long processes. Like reticular cells, macrophages establish close physical associations with maturing blood cells, notably erythroid cells. However, reticular cells are physically associated with most blood cell precursors, and in particular granulocyte precursors. In the electron microscope, it was noted that reticular cells have gap junctions connecting their cell processes to neighbouring cells and to endosteal lining cells and that they contain bundles of thin filaments, which suggest an ability to retract (a property claimed to be related to uncovering of portions of the sinus wall and to transendothelial blood cell migration). Reticular cells are thought to produce 'reticulin' (collagen type III-rich) fibres, which represent the main (though scanty) fibrillar component of the marrow extracellular matrix, and this has led to the suggestion that reticular cells are fibroblastic in nature. Reticular cells are candidates for *in vivo* counterparts, albeit in a broad sense, of fibroblast-like cells that grow in culture from marrow stroma.

Adipocytes are a major feature of post-natal marrow stroma. When seen in a two-dimensional section, adipocytes appear to be distributed haphazardly across the marrow space. They are part of the sinusoidal adventitia, and are in contact with the outer surface of the endothelium, similar to 'adventitial reticular cells'. Adipocytes are lined by a basement membrane-like material that is immunoreactive for collagen type IV and laminin. Because of their large size, adipocytes encroach upon and compress the wall of the sinusoids, and tend to reduce transendothelial migration of marrow and blood-borne cells.

'Reticulum' cells, characterised by long cytoplasmic extensions and plasma membrane-associated ALP activity, which enabled them to be distinguished from macrophages, were described in rodent and murine marrow by Westen and Bainton (1979). Using ALP cytochemistry on specially prepared, 1–2 μm thick, plastic-embedded tissue sections, these cells have now been detected in the marrow of many species including human (Bianco *et al.*, 1988; Bianco & Bonucci, 1991) although variations occur in their prominence and numbers. In these samples, individual ALP+stromal cells appear mostly as slender filaments outlined by reaction product, whereas in thicker (10 μm or more) sections, the pattern of distribution of reaction product resembles a reticulin network. Whether all 'reticulum' cells express ALP activity at all times remains to be determined. ALP activity and immunoreactivity for NGF receptors codistribute in tissue sections (Cattoretti *et al.*, 1993). The patterns of immunolabelling observed for collagen type I and III and osteonectin are also reminiscent of the distribution of ALP activity, although an actual codistribution has not been demonstrated formally.

Studies with confocal microscopy (Bianco & Boyde, 1993) have indicated that the ALP+cells described by Westen and Bainton are the same cell type as the adventitial reticular cells of earlier EM studies, and herein after, they will be referred to as ALP+reticular cells. The regular association of these cells with arteriolar and sinusoidal walls, which can be missed in 1–2 μm thick histological sections, is readily appreciated using 3D confocal microscopy. A close association between haematopoietic cells and niches made out of the ALP+processes of reticular cells was also demonstrated using the same technique.

Using ALP activity as a marker of reticular cells, changes in this cell type in human bone marrow can be studied. In the normal state, ALP+reticular cells are found throughout the marrow and also near to bone surfaces. It is well known that those near bone surfaces have osteogenic potential and, in states of increased bone-forming activity, such as hyperparathyroidism (Bianco & Bonucci, 1991), there are increased numbers and unusual spatial arrangements of these cells. In the early stages of osteosclerosis (excess formation of bone within marrow) due to cancer metastasis to bone, the earliest foci of osteoid are deposited within sites of increased density of ALP+cells in the adventitia of marrow sinusoids. ALP+cells appear to provide 'docking space' and physical support to newly forming clusters of haematopoietic cells following bone marrow transplantation.

Taken together, observations made on developing, normal post-natal, and pathological marrow suggest that ALP+reticular cells in the marrow environment originate as part of a primarily osteogenic cell population, but may be recruited to serve one or other of two highly differentiated functions: either to contribute to a stromal microenvironment for the support of haematopoiesis, or to grow and mature into bone-forming cells.

Adipocytes

Marrow adipogenesis is a post-natal event (Tavassoli & Yoffey, 1983). At birth, all available space within bone is occupied by haematopoietic cells and their supportive, non-adipose, stroma. As bone grows, space available for haematopoiesis exceeds the total volume of haematopoietic cells and this extra space becomes occupied by newly developed adipocytes. This process is not uniform across the skeleton, and the reason for its diversity is unclear. In the appendicular skeleton, adipogenesis progresses from the mid-diaphyseal marrow, where the first adipocytes appear at birth, towards the epiphyses. The process continues until, at skeletal maturity, the whole of the diaphyseal marrow cavity is normally filled with adipose tissue (yellow marrow) and haematopoietically active marrow (red marrow) is restricted to the cancellous bone of metaphyses and epiphyses (Fig. 2.3) (Neumann's law; Neumann, 1882). In the axial skeleton, the marrow remains haematopoietically active (red) throughout life, yet the number of adipocytes per unit volume steadily increases. It is estimated that the proportion of marrow volume occupied by adipocytes shifts in the iliac crest from approximately 30% in early adulthood to 60% or more at the age of 60.

Growth and ageing of the skeleton is thus one determinant of adipogenesis in the marrow (Fig. 2.3). Also a contributing factor is the total haematopoietic cell mass. At any skeletal site in the post-natal marrow, if the volume of haematopoietic tissue expands, it does so at the expense of adipocytes. If extra space for haematopoiesis is needed during childhood, when there is no adipose tissue present, it will be at the expense of bone, resulting in major changes in bone structure. After bone growth is completed, haematopoiesis competes for space with adipose (yellow) marrow. If the bone marrow has to compensate for increased demand of blood cells (as in chronic anaemia, for example), it will expand and occupy extra space given up by yellow marrow. A markedly hypercellular marrow, such as a leukaemic or hyperplastic marrow, does not contain adipocytes, and is packed with haematopoietic or leukaemic cells. A severely hypoplastic or a hypocellular marrow, on the other hand, is extensively filled with adipocytes. The volume of marrow occupied by adipocytes and haematopoietic cells, respectively, are inversely related to one another.

Changes in the number of adipose cells that occur as a function of changes in total haematopoetic cell mass, are mirrored by changes in numbers of ALP+reticular cells. A hypercellular marrow contains increased numbers of these cells, whereas if fatty involution has taken place and the marrow is hypocellular, no ALP+reticular cells are found. The content of ALP+retic-

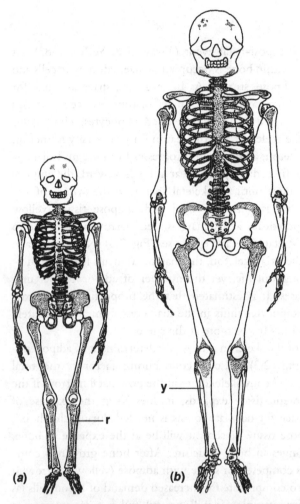

Fig. 2.3. Red (haematopoietically active) and yellow (adipose) marrow in the human skeleton (Neumann's law). (*a*) At birth, all marrow cavities in all bones are occupied by haematopoietically active red marrow (r). (*b*) At skeletal maturity, red marrow is restricted to the axial skeleton. In limb bones, the diaphyseal marrow cavities are occupied by yellow adipose marrow (y) and the cancellous or trabecular bone of epiphyses and metaphyses remains associated with red marrow (r). The age-dependent change from red to yellow marrow is reversible in some pathological conditions, see text.

ular cells in the marrow can be estimated in histological sections, by measuring the total linear extension of ALP-reactive cell processes, compared with the total area of marrow. The value for ALP density (ALP-d) thus obtained, is a linear function of the space occupied by haematopoietic cells

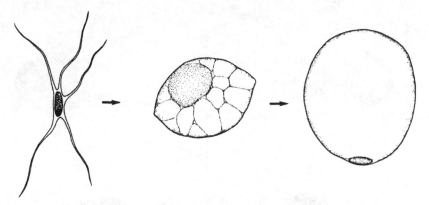

Fig. 2.4. Diagram showing the conversion of reticular cells to adipocytes.
Multivacuolar pre-adipocytes represent the intermediate step in this conversion, and
are easily detected in the human post-natal marrow in association with any instance
of rapid depletion of haematopoietic cells. Preadipocytes express ALP activity, which
is lost during their further maturation to adipocytes. This explains why no ALP+
stromal cells are found in a volume of marrow that has undergone extensive 'fatty'
involution.

and is inversely correlated with that occupied by adipocytes (Bianco *et al.*,
1993*a*).

Early electron microscopic studies on age-related adipogenesis had suggested that 'adventitial reticular cells' could become adipocytes, by progressive accumulation of lipid (Fig. 2.4) (Weiss & Sakai, 1984). In the human
marrow, in patients with acute leukaemia, following administration of anti-
leukaemic, myeloablative drugs, an increase in adipogenesis parallels rapid
reduction in the total haematopoietic cell mass (Bianco *et al.*, 1988).
Haematopoietic cell reduction is accompanied by the appearance of pre-
adipocytes containing lipid vacuoles of variable size. Like ALP+reticular
cells, pre-adipocytes express plasma membrane associated ALP activity, but
this is gradually lost as they progress towards terminal stages of lipid
accumulation. This suggests that ALP+reticular cells in the adult bone
marrow can become adipocytes, once released from their associations with
haematopoietic cells.

ALP+reticular cells and adipocytes are not to be considered as separate
lineages but rather as alternative phenotypes into which stromal cells in the
post-natal marrow may modulate depending on environmental influences
(Bianco *et al.*,1988). Similarly, adipogenesis and osteogenesis are alternative,
but not irreversible, pathways of differentiation. This is suggested by recent
experimental evidence, showing that marrow adipocytes can be switched

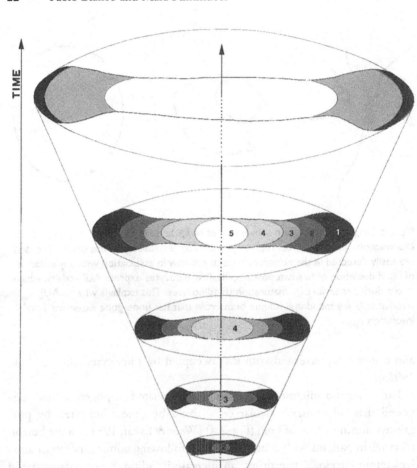

Fig. 2.5. The diagram shows schematic longitudinal sections of a human humerus at different times, from chondrogenesis through skeletal maturity. From bottom to top, is shown a representation of a humerus from the following stages: 10 mm embryo, 27 mm embryo, 50 mm embryo, 52 mm embryo, birth, adult. Subsequent phases of endochondral osteogenesis (chondrogenesis, cartilage hypertrophy, formation of a primary bony spongiosa with pre-haematopoietic stroma), myelogenesis, and adipose involution of the marrow as per Neumann's law (that is, pre-natal and post-natal changes) are viewed as part of a single multi-step developmental process. Each phase has been identified by the appearance of a novel phenotype of the stromal system, and each phenotype has been numbered consecutively. 1, cartilage; 2, hypertrophic cartilage; 3, primary bone spongiosa and pre-haematopoietic marrow; 4, haemato-poietic marrow; 5, marrow adipocytes. Each new phase begins at the mid-diaphysis of

towards osteogenic differentiation in culture (Bennett *et al.*, 1991; Beresford *et al.*, 1992). These data, and the *in vivo* observations discussed above, suggest that the divergence of stromal lineages does not occur necessarily at the stem cell level. If true, this would clearly have important implications for under-standing the manner in which lineage hierarchy is laid out within the stromal system. The stromal system was originally conceived as a number of separate lineages emanating from a common stem cell by analogy with the haematopoietic system (Owen, 1985). Osteoblasts, reticular cells and adipocytes would be mature, differentiated, phenotypes of their respective lineages. However the lineages of the bone marrow stromal system, it appears, are able to switch from one to another at points downstream to the putative stromal stem cell (see also Owen, Chapter 1). This plasticity of the differentiated cells is the cellular basis for changes occurring as a function of time and disease.

Bone and marrow and fat – and time

Tissues and cells that represent the main different phenotypes of the stromal system, when viewed as part of an organ, and of its changes in development, growth, and ageing, display a remarkable level of organisation, in space and time (Fig. 2.5). They appear in a temporal sequence, and this is reproduced at each time point in a spatial sequence along the direction of bone growth. Time dictates the ranking of lineages in the stromal system. Chondrogenesis, osteogenesis, pre-haematopoietic stroma, myelogenesis, and adipogenesis are subsequent eras in the history of a bone and marrow organ (Fig. 2.5). Adipogenesis is the sole, entirely post-natal era in its history.

Adipose involution represents the future of substantial portions of haematopoietic marrow as a time-dependent process enforced by a physio-logical law. Meunier *et al.* (1971) reported increased numbers of adipocytes in the bone marrow associated with osteoporotic bone, and involutional

the skeletal rudiment. As a result of the dominant direction of bone growth, each phase expands towards the epiphyses. In a section of bone at birth, the different tissues of the stromal system are encountered in a spatial sequence along the direction of bone growth. The same sequence is encountered in an imaginary section in time (arrow in the axis of the cone). In the top diagram of a normal adult, the primary spongiosa of cancellous bone and red marrow lie next to epiphyseal (articular) cartilage. The diaphyseal marrow has undergone adipose 'involution' as per Neumann's law. A further diagram of an ageing bone (not shown) would see the adipose marrow expand at the expense of the cancellous bone. This is what happens in involutional, senile, osteoporosis.

bone loss in the elderly consists of the substitution of cancellous bone with adipose tissue. Stated as a paradox, involutional osteoporosis is the process whereby bone, with time, turns into fat. This process appears to be related to the ageing of bone as an organ, and can be seen as the ultimate horizon of bone development. Studies designed to understand this process, and hopefully to provide a cure for its pathological expressions, must take into account two important characteristics of the marrow stromal system in which its cellular basis resides: (a) the plasticity of phenotypes and the continuing influence of microenvironmental cues on their expression, and (b) the bearing of biological time (growth, ageing) on the evolution of the system.

Acknowledgement

We thank Dr Diego De Merich for skilled help with the illustrations.

References

Ascenzi, A. (1976). Physiological relationship and pathological interferences between bone tissue and marrow. In *The Biochemistry and Physiology of Bone*, ed. G. Bourne, vol IV, pp. 403–44. New York: Academic Press.

Bennett, J.H., Joyner, C.J., Triffitt, J.T. & Owen, M.E. (1991). Adipocytic cells cultured from marrow have osteogenic potential. *J. Cell Sci.*, **99**, 131–9.

Beresford, J.N., Bennett, J.H., Devlin, C., Leboy, P.S. & Owen, M.E. (1992). Evidence for an inverse relationship between the differentiation of adipocytic and osteogenic cells in rat marrow stromal cell cultures. *J. Cell Sci.*, **102**, 341–51.

Bianco, P. & Bonucci, E. (1991). Endosteal surfaces in hyperparathyroidism: an enzyme cytochemical study on low temperature-processed, glycol methacrylate embedded bone biopsies. *Virchow's Arch. A Pathol. Anat. Histopath.*, **419**, 425–31.

Bianco, P & Boyde, A. (1993). Confocal images of marrow stromal (Westen–Bainton) cells. *Histochemistry*, **100**, 93–9.

Bianco, P., Constantini, M., Dearden, L.C. & Bonucci, E. (1988). Alkaline phosphatase positive precursors of adipocytes in the human bone marrow. *Br. J. Haemat.*, **68**, 401–3.

Bianco, P., Bradbeer, J.N., Riminucci, M. & Boyde, A. (1993a). Marrow stromal (Westen–Bainton) cells: identification, morphometry, confocal imaging and changes in disease. *Bone*, **14**, 315–20.

Bianco, P., Riminucci, M., Bonucci, E., Termine, J.D. & Gehron-Robey, P. (1993b). Bone sialoprotein (BSP) secretion and osteoblast differentiation: relationship to bromodeoxyuridine incorporation, alkaline phosphatase and matrix deposition. *J. Histochem. Cytochem.*, **41**, 183–91.

Cattoretti, G., Schiro, R., Orazi, A., Soligo, D. & Colombo, M.P. (1993). Bone marrow stroma in humans: anti-nerve growth factor receptor antibodies selectively stain reticular cells *in vivo* and in *vitro*. *Blood*, **81**, 1726–38.

Gimble, J.M., Wanker, F., Wang, C.S., Bass, H., Wu, X., Kelly, K., Yancopoulos, G.D. & Hill, M.R. (1994). Regulation of bone marrow stromal cell differentiation by cytokines whose receptors share the gp130 protein. *J. Cell Biochem.*, **54**, 122–33.

Meunier, P., Aaron, J., Edouard, C. & Vignon, G. (1971). Osteoporosis and the replacement of cell populations of the marrow by adipose tissue. A quantitative study of 84 iliac bone biopsies. *Clin. Orthop.*, **80**, 147–54.

Neumann, E. (1882). Das Gesetz Verbreitung des gelben und roten Markes in den Extremitatenknochen. *Centr. Med. Wiss.*, **20**, 321–3.

Owen, M.E. (1985). Lineage of osteogenic cells and their relationship to the stromal system. In *Bone and Mineral Research*, ed. W.A. Peck, vol III, pp. 1–25. Amsterdam: Elsevier.

Rouleau, M., Mitchell, J. & Goltzmann, D. (1988). *In vivo* distribution of parathyroid receptors in bone: evidence that a predominant osseous target is not the mature osteoblast. *Endocrinology*, **123**, 187–91.

Streeter, G. (1949). Developmental horizons in human embryos. *Contrib. Embryol., Carnegie Inst. Washington*, **33**, 151–67.

Tavassoli, M. & Yoffey, Y.M. (1983). *Bone Marrow: Structure and Function*. New York: Alan R. Liss.

Weiss, L. (1976). The haematopoietic microenvironment of bone marrow: an ultrastructural study of the stroma in rats. *Anat. Rec.*, **186**, 161–84.

Weiss, L. & Sakai, H. (1984). The hematopoietic stroma. *Am. J. Anat.*, **170**, 447–63.

Westen, H. & Bainton, D.F. (1979). Association of alkaline phosphatase positive reticulum cells in the bone marrow with granulocytic precursors *J. Exp. Med.*, **150**, 919–37.

3

Isolation, purification and *in vitro* manipulation of human bone marrow stromal precursor cells

Stan Gronthos, Stephen E. Graves and Paul J. Simmons

Introduction

Precursors of the marrow stromal system

The cell types comprising the stromal tissue of the bone marrow (BM) include reticular cells, smooth muscle cells, adipocytes, osteoblasts and various different populations of vascular endothelial cells (Lichtman, 1981; Tavassoli & Friedenstein, 1983; Dexter *et al.*, 1984; Allen, Dexter & Simmons, 1990). A similarly diverse population of stromal cells develops *in vitro* when BM cells are explanted under appropriate conditions, as originally described by Dexter and colleagues. This well-documented heterogeneity of marrow stromal cells has complicated attempts to characterise the biological properties of each cellular component, a problem compounded by the paucity of monoclonal antibody reagents which might facilitate precise identification and isolation of each cell type.

Studies in rodents and humans have shown that the bone marrow stroma has the ability to regenerate either following physical disruption of the marrow cavity or following high dose chemotherapy or radiation (Patt & Maloney, 1975; Simmons *et al.*, 1987; Testa, Hendry & Molineux, 1988). Given the heterogeneity of the stromal cell population within the bone marrow microenvironment (BMME), it is not known whether all of the different stromal cell lineages have the capacity for self-renewal or, alternatively, whether each stromal cell type arises from the proliferation and differentiation of a common stromal stem cell pool. Putative BM stromal precursor cells (SPC) have been identified in a number of species, including humans, by their ability to generate colonies of cells morphologically resembling fibroblasts when single cell suspensions of BM are explanted at appropriate densities in liquid culture (Fig. 3.1). The clonogenic stromal pro-

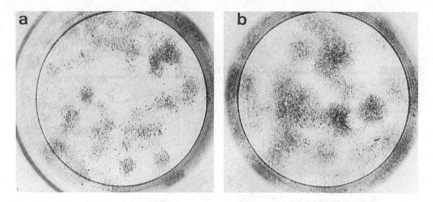

Fig. 3.1. Representative photographs of culture dishes showing the proliferation of BM CFU-F, cultured in: a, standard CFU-F growth medium supplemented with 20% FCS and b, SDM supplemented with EGF+PDGF, 10 ng/ml.

genitor cells responsible for colony growth under these conditions are referred to as fibroblast colony-forming cells (CFU-F) (Friedenstein, Chailakhyan & Lalykina, 1970; Castro-Malaspina *et al.*, 1980). In this chapter, the term CFU-F will be reserved for stromal precursor cells which demonstrate clonogenic growth *in vitro* under appropriate assay conditions. The acronym SPC will be used to encompass all clonogenic stromal precursors irrespective of the method of assay (*in vitro* or *in vivo*) taking into consideration the possible existence of SPC in the BM which may not be detected by current *in vitro* assays and which only demonstrate their potential in an *in vivo* setting.

CFU-F were originally identified in rodent BM by Friedenstein and colleagues. Results obtained using a variety of different techniques collectively support the conclusion that each colony is derived from a single cell (Friedenstein *et al.*, 1970; Castro-Malaspina *et al.*, 1980; Perkins & Fleischman, 1990). *In vivo* CFU-F are almost entirely in a non-cycling state as demonstrated by [³H] thymidine labelling in rodents and by means of the *in vitro* thymidine suicide technique in adult human BM (Castro-Malaspina *et al.*, 1980).

CFU-F colonies derived from the BM of virtually all species examined, including humans, are heterogeneous in size and morphology (Fig. 3.1), prompting the suggestion that they originate from clonogenic progenitors at various stages of differentiation (Owen, Chapter 1). A proportion of the colonies are large in size and demonstrate extensive replating potential after passaging (Friedenstein *et al.*, 1987). The high proliferative potential of a proportion of the clonogenic population led Friedenstein to propose that

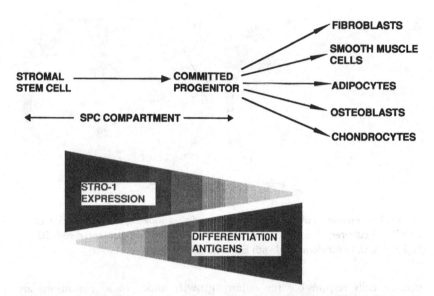

Fig. 3.2. The stromal stem cell hypothesis modified (see also Owen, Chapter 1). The expression of STRO-1, which is proposed as a developmental marker, is high initially but declines as the SPC proliferate and differentiate.

within the CFU-F compartment are precursor cells with stem cell character-istics. Consistent with this interpretation are the results of a study in which the progeny of individual CFU-F cultured from mouse BM were implanted beneath the renal capsule of syngeneic hosts. Of the large colonies implanted, a minor proportion (approximately 15%) produced a BM organ containing the full spectrum of stromal cell types, including osteoblasts. A further 15% produced bone tissue only while the remainder formed soft connective tissue only or failed to give rise to any tissue (Friedenstein, 1980). Other investigators have shown the induction of adipogenesis in a propor-tion of BM-derived CFU-F when cultured in the presence of hydro-cortisone (Bennett *et al.*, 1991). In that same study it was also shown that approximately 40% of the adipocytic colonies were capable of forming bone when transplanted *in vivo* within diffusion chambers.

In the early 1980s based on these studies, the stromal stem cell hypothesis was proposed in which, by analogy with the organisation of the haemopoi-etic system, there exists a hierarchy of cellular differentiation supported at its apex by a small number of multipotential, self-renewing stromal stem cells (Fig. 3.2) (references in Owen, Chapter 1). Thus, CFU-F which yielded BM organs were hypothesised to be stromal stem cells, while those that gave rise to bone or soft connective tissue only were proposed to be of more restricted developmental potential. It must be emphasised that, whilst these data provide strong circumstantial evidence for the existence of stromal stem cells, definitive proof of the hypothesis will require more rigorous studies involving, for example, transplantation of uniquely (retrovirally) marked SPC. However, this approach is limited to putative SPC which can be cloned *in vitro* as CFU-F.

This chapter explores current ability to isolate and manipulate the growth of human BM SPC *in vitro*. Potential clinical applications of culture-expanded SPC, which it is believed may provide the basis for a number of novel and valuable cellular therapies, include the replacement of stroma damaged following high dose chemotherapy and radiation in cancer patients receiving bone marrow transplants, the generation of osteogenic cells for re-implantation into areas of bone loss and the use of stromal stem cells as vehicles for gene therapy (for review, see Gronthos & Simmons, 1996).

Characterisation of stromal precursor cells (SPC)

Cell surface antigens expressed by human BM SPC

CFU-F, an *in vitro* compartment of SPC, have been isolated from a variety of different haemopoietic tissues including spleen, lymph node, thymus and bone marrow of adult mice, rats, guinea pigs, rabbits and dogs (Friedenstein, 1980). CFU-F have also been derived from fetal mouse liver, spleen, and bone marrow (van den Huevel, 1987). In humans, CFU-F have been isolated from both adult and fetal bone marrow and fetal liver (Huang & Terstappen, 1992; Castro-Malaspina et al., 1980). No CFU-F have been detected in the steady state peripheral blood of adult humans or in fetal human cord blood (S. Gronthos & P. J. Simmons, unpublished observations).

Terstappen and colleagues have conducted an analysis of the antigens expressed by SPC in fetal BM (Huang & Terstappen, 1992; Waller et al., 1995). SPC were shown to express the CD34 antigen at high level but were distinguished from the bulk of haemopoietic progenitors by their lack of CD38 and HLA-DR. Based on FACS studies it was further claimed that within this $CD34^+CD38^-HLA\text{-}DR^-$ subpopulation were so-called 'common stem cells' with the potential to develop into both haemopoietic and stromal cell progeny (Huang & Terstappen, 1992). This conclusion was subsequently retracted, however, following their demonstration that haemopoietic progenitors were contained within a $CD50^+$ (ICAM-3) subset of the $CD34^+$ $CD38^-HLA\text{-}DR^-$ cells while SPC were restricted to a $CD50^-$ subset (Waller et al., 1995).

A monoclonal antibody (Mab), STRO-1 has previously been described which identifies essentially all assayable CFU-F in aspirates of adult human BM (Simmons & Torok-Strob, 1991a). STRO-1 was developed following immunisation of mice with CD34$^+$ BM cells. Consistent with the above cited observations of Terstappen et al. in fetal human BM, antibodies to CD34 were subsequently shown to bind to CFU-F in adult BM (Simmons & Torok-Storb, 1991b). However, unlike the high level of CD34 antigen expression characteristic of fetal BM SPC, adult BM SPC express CD34 at considerably lower levels, at least ten-fold less than that exhibited by haemopoietic progenitors in the same sample. In addition, not all CFU-F are recovered in the CD34$^+$ fraction which might reflect either inefficient capture of SPC as a result of low CD34 antigen density or heterogeneity of CD34 expression within the SPC compartment (Simmons & Torok-Storb, 1991b, Waller et al., 1995). For these reasons, therefore, selection of SPC based on the use of CD34 antibody does not represent the method of choice for isolating these cells from adult human BM.

STRO-1+BM mononuclear cells (BMMNC) are a heterogeneous population comprising predominantly (approximately 90%) late stage glycophorin A+erythroblasts and a small proportion of CD19$^+$ B-cells. CFU-F are restricted to the minor population of STRO-1+cells, which are characterised by a high level of STRO-1 antigen expression and a lack of glycophorin A (Simmons & Torok-Storb, 1991a). The expression of a wide range of cell surface antigens by BM CFU-F has been examined systematically using two-colour fluorescence activated cell sorting (FACS) (Simmons et al., 1994; Gronthos and Simmons, 1995). A number of antigens were shown to be expressed by essentially all CFU-F. These include the adhesion molecules Thy-1 (CDw90) and VCAM-1 (CD106) of the immunoglobulin superfamily and various β1 (CD29) integrin family members α2β1 (CD49b/CD29) and α5β1 (CD49e/CD29). CFU-F were further characterised both by the expression of endopeptidases CD10 and CD13 and the growth factor receptors for platelet derived growth factor (PDGF), epidermal growth factor (EGF), insulin-like growth factor-1 (IGF-1), and nerve growth factor (NGF). Antigens restricted to haemopoietic cells (CD3, CD19, CD33, CD38, CD45, glycophorin A and HLA-DR) were found not to be expressed at detectable levels on adult human BM CFU-F (Simmons et al., 1994). This is in accord with the recent data of Terstappen and colleagues regarding the antigenic phenotype of human fetal BM SPC (Waller et al., 1995).

The Mab STRO-1 has, in many respects, proved to be an ideal reagent for isolation of SPC from adult BM. Importantly, STRO-1 demonstrates no detectable binding to haemopoietic progenitors (CFU-GM, BFU-E, BFU-Meg, CFU-GEMM) nor to their precursors (pre-CFU) (Simmons & Torok-Storb, 1991a; S. Gronthos and P.J. Simmons, unpublished observations), thus facilitating a clean separation between SPC and haemopoietic progenitors in adult BM. Moreover, the antigen(s) identified by STRO-1 is expressed at particularly high copy number by CFU-F (S. Gronthos & P.J. Simmons, unpublished observations) which may, in part, account for the efficiency of CFU-F isolation with STRO-1.

Assay of CFU-F from aspirates of human BM yields colony numbers which are generally in the range 1 per 5000–10000 mononuclear cells (Castro-Malaspina et al., 1980; Simmons & Torok-Storb, 1991a,b; Waller et al., 1995). The rarity of these clonogenic SPC is a major limitation to their study, a problem compounded until recently by a lack of efficient enrichment procedures. FACS provides an ideal means for isolating enriched populations of STRO-1+cells for laboratory studies (Simmons & Torok-Storb,1991a,b; Gronthos et al., 1994) but without an efficient pre-enrichment step is not practical for larger (clinical) scale isolations.

The developmental potential of human BM SPC

Very little is currently known regarding the developmental potential of SPC derived from adult human BM, but this remains a major objective of research efforts in this field. Single SPC isolated by FACS from fetal human BM were claimed to have the potential for differentiation into multiple stromal cell lineages including bone and cartilage (Huang & Terstappen, 1992). While an important study, these data must, however, be interpreted with caution since for several stromal elements (particularly the osteogenic and chondrogenic lineages), cell type specific, immunohistologically defined markers of differentiation were not employed, their identification being based on morphological criteria only.

Adult BM cells sorted on the basis of STRO-1 expression have previously been shown to develop into an adherent stromal layer when grown under long-term culture conditions with an increased capacity to support haemopoiesis when compared with the stroma derived from unfractionated BM. The adherent layers derived from STRO-1+BM consist of a number of phenotypically distinct stromal cell types including fibroblasts, smooth muscle cells and adipocytes (Simmons & Torok-Storb, 1991a).

Isolation of BM SPC

The following methods are used routinely in the authors' laboratory for the pre-enrichment of CFU-F from adult human bone marrow prior to their culture.

Materials

Polypropylene tubes, 4 ml, 12 ml, 50 ml (Nunc, Falcon, Becton Dickinson, Linkon Park, NJ)
Preservative-free heparin 300 U (David Bull Laboratories, Sydney, NSW, Australia)
Hank's buffered saline solution (HBSS)
Fetal calf serum (FCS) (GIBCO BRL, Victoria, Australia)
HEPES 10 mM, pH 7.35 (GIBCO BRL)
Ficoll 1.077 g/ml; (Lymphoprep; Nycomed, Oslo, Norway)
Bovine serum albumin (BSA) 1% Cohn fraction V (Sigma)
Normal human AB serum 1%
Trypsin 0.5%
EDTA 5 mM
PBS Ca^{2+} and Mg^{2+} free (GIBCO, BRL)
Formalin 1% in PBS
Glucose 0.02 g/ml
Sodium azide 0.01%
Falcon cell strainer (Becton Dickinson Labware, Franklin Lakes, NJ)
100 μl streptavidin microbeads (Miltenyi Biotec, Bergisch Gladbach, FRG)
Goat anti-mouse IgM μ-chain specific fluorescein isothyiocyanate (FITC) and goat anti-mouse IgG γ-chain specific phycoerythrin (PE) (Southern Biotechnology Associates, Birmingham, AL)
Streptavidin–FITC conjugate (Caltag Laboratories, San Francisco, CA)
FACStar^{PLUS} flow cytometer (Becton Dickinson, Sunnyvale, CA)

Preparation of bone marrow samples

Bone marrow samples are aspirated from the posterior iliac crest or sternum of normal adult volunteers and marrow allogeneic donors after informed consent and ethical approval. The BM is collected into 50 ml polypropylene tubes containing 300 units of preservative-free heparin and mixed. The bone marrow cells are then diluted (1:1) in HBSS supplemented with 5% FCS and 10mM HEPES, pH 7.35. Low density bone marrow mononuclear cells

(BMMNC) are isolated by density gradient centrifugation over Ficoll (1.077 g/ml) at 400 *g* for 30 minutes. Following this, the BMMNC layers are carefully removed by pipette then washed twice in HBSS+5% FCS at 400 *g* for 5 minutes at 4 °C and maintained on ice, prior to immunolabelling, flow cytometric analysis and cell culture.

Isolation of STRO-1+cells from BMMNC by indirect fluorescence-activated cell sorting (FACS)

Single label

The following protocol is for the isolation of CFU-F from adult BM by FACS using STRO-1 alone. BMMNC are resuspended in blocking buffer (HBSS+10 mM Hepes, 5% FCS, 1% normal human AB serum and 1% BSA), and incubated for 30 minutes on ice to block non-specific binding to Fc receptors. Approximately 1×10^7 cells (for cell sorting) and 2×10^5 cells (for phenotypic analysis) are pelleted in 12 ml and 4 ml polypropylene tubes respectively and resuspended in 50–200 μl of the undiluted monoclonal antibody STRO-1 supernatant for 1 hour on ice with occasional mixing. In a separate tube the control cells ($2 \times 10^5 - 5 \times 10^5$) are reacted with the mouse monoclonal isotype IgM negative control antibody, 1A6.12 (kindly donated by Dr. L.K. Ashman; Department of Haematology, IMVS, Adelaide, SA, Australia) under the same conditions. The cells are then washed in HBSS+5% FCS and the secondary antibody (1/30 dilution) of μ-chain specific, goat anti-mouse IgM FITC added in a final volume of 100 μl. The cells are then incubated for 45 minutes on ice, and subsequently washed twice in HBSS+5% FCS. STRO-1 positivity is defined as the level of fluorescence greater than 99% of the isotype matched control antibody 1A6.12. Cells for flow cytometric analysis only are resuspended in 500 μl of FACS FIX solution (1% formalin in PBS+0.02 g/ml glucose+0.01% sodium azide) and can be stored at 4 °C for up to several weeks until required.

Dual Label

In order to increase the incidence of CFU-F in the STRO-1 selected fraction, additional purification steps are required. Recent studies have demonstrated that CFU-F incidences in the range of 1 per 10–20 cells can be obtained from aspirates of normal human adult BM by two-colour FACS using STRO-1 together with appropriate combinations of several different cell surface molecules described above. This represents up to a 1000-fold enrichment over starting SPC numbers (S. Gronthos and P.J. Simmons, manuscript in preparation).

The following method describes the purification of CFU-F from adult human BM by two-colour FACS. BMMNC cells $(1 \times 10^6 – 10 \times 10^6)$ are resuspended in a 12 ml polypropylene tube in 5 ml of blocking buffer and incubated for 30 minutes on ice. Following this, the cells are pelleted and resuspended in 200 μl of primary antibody cocktail for 1 hour on ice with occasional mixing. The primary antibody cocktail consists of saturating concentrations of mouse IgM monoclonal antibody STRO-1 (1: 1 dilution of hybridoma supernatant) and either one of the mouse IgG monoclonal antibodies (20 μg/ml) shown in previous studies to bind to CFU-F (Simmons *et al.*, 1994; Gronthos & Simmons, 1995). Several IgG1 Mabs when used in combination with STRO-1 for two-colour FACS have produced high levels of CFU-F enrichment of between 1000- and 3000-fold in comparison to unfractionated BM (S. Gronthos and P.J. Simmons unpublished observations). These include RMAC11 (anti-CD49b/CD29; Dr G. Russ and Dr R. Faull, Queen Elizabeth Hospital, Adelaide, SA, Australia), PHM2 (anti-CD49e/CD29; kindly donated by Prof. R.A. Aitkins, Monash Medical Centre, Melbourne, Vict. Australia), 6G10 (anti-VCAM-1; kindly donated by Dr B. Masinovsky, ICOS Corp., Seattle, WA) and F15.42 (anti-Thy-1; kindly donated by Dr D. Hart, Department of Haematology, Christchurch Hospital, New Zealand). The mouse monoclonal IgM isotype (1A6.12) and IgG1 isotype (3D3) negative control antibodies (kindly donated by Dr L.K. Ashman; Department of Haematology, IMVS, Adelaide, SA, Australia) are used under identical conditions. Control tubes should also be set up to facilitate electronic colour compensation on the cell sorter or analyser by staining cells with (a) STRO-1 and the IgG1 negative control antibody (3D3) and (b) IgG1 antibody specific to the cell surface marker of interest and the IgM control antibody (1A6.12). The cells are then washed in HBSS+5% FCS and second labels are added in a final volume of 100 μl. These consist of goat anti-mouse IgM μ-chain specific FITC (1/30) and goat anti-mouse IgG γ-chain specific PE (1/50). Following an incubation period of 45 minutes on ice, the cells are washed twice in HBSS+5% FCS and then resuspended to approximately 10^7 cells/ml prior to being sorted using a FACStar^PLUS flow cytometer. Positivity for each fluorochrome is defined as the level of fluorescence greater than 99% of the appropriate isotype matched control antibodies.

Analysis of the progeny of BM SPC

The following method describes the isolation and flow cytometric analysis of cultured stromal cells derived from STRO-1+BMMNC. To obtain single cell suspensions, stromal cultures are first rinsed once in PBS and then treated with 0.5% trypsin+5 mM EDTA in PBS at 37 °C for 5–10 minutes. The trypsinised cells are then diluted and washed in cold HBSS+5% FCS to inactivate the trypsin. Prior to analysis, the cell suspension is passed through a Falcon cell strainer to remove any remaining cell aggregates. The cells are then maintained on ice in blocking buffer for 20 minutes before being recovered by centrifugation and resuspended in primary antibody. One-colour and two-colour FACS is performed according to protocols given on p. 33 and p. 34, respectively.

Isolation of STRO-I +cells by magnetic-activated cell sorting (MACS)

The magnetic-activated cell sorting (MACS) procedure described by Miltenyi and colleagues (Miltenyi *et al.*, 1990) provides a particularly effective means for the large-scale isolation of SPC from BMMNC (Gronthos & Simmons, 1995).

BMMNC (1×10^8 cells) are suspended in 1 ml of STRO-1 supernatant in a 12 ml polypropylene tube for 60 minutes on ice with occasional mixing. The cells are then washed twice in HBSS supplemented with 1% BSA and resuspended in 1 ml of buffer containing a 1/50 dilution of biotinylated μ-chain specific goat anti-mouse IgM for 45 minutes on ice. The cells are then washed twice in MACS buffer (comprising single strength Ca^{2+} and Mg^{2+} free PBS supplemented with 1% BSA, 5mM EDTA and 0.01% sodium azide) and degassed under vacuum prior to use in order to eliminate air bubbles which might block the magnetic column matrix, and recovered by centrifugation prior to resuspension in 900 μl of MACS buffer to which 100 μl of streptavidin microbeads is added. The cells are incubated for 15 minutes on ice with occasional mixing after which streptavidin-FITC conjugate (1/50) is added directly to the suspension for an additional 5 minutes. The cells are then washed twice in MACS buffer, and a small aliquot of the cell suspension removed for flow cytometric analysis (Fig. 3.3). The remainder of the cell suspension is then applied to a Mini MACS magnetic column (column capacity 10^7 cells, Miltenyi Biotec) prepared as follows. MACS buffer (500 μl) is first passed through the magnetised column to remove the protective coating of the column matrix. The cells in MACS buffer (about 1 ml) are then applied to the column at a flow rate of approximately 0.3

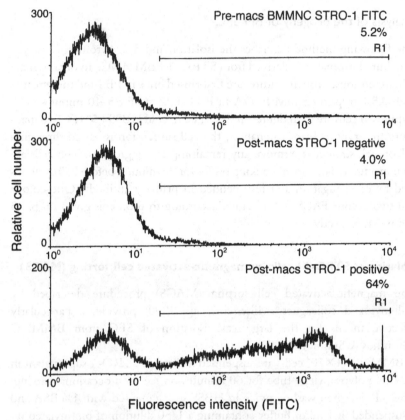

Fig. 3.3. Frequency histograms representing the flow cytometric analysis of BMMNC sorted by mini MACS, on the basis of STRO-1 expression: unsorted BMMNC fraction (top); STRO-1 − cell fraction (centre); STRO-1 + cell fraction (bottom). Each histogram is based on 10^4 events collected as list mode data.

ml/minute. The column flow through, containing the negative cell fraction (STRO-1- cells), is collected in a 4 ml polypropylene tube and placed on ice. The column is then washed with an additional 1 ml of MACS buffer. To recover the bound, STRO-1+cells the column is first removed from the magnet and then flushed with 2 ml of MACS buffer into a sterile tube using the column plunger. An aliquot from each of the STRO-1+and STRO-1- cell populations is analysed by flow cytometry in order to determine the purity and recovery of cells in each fraction (Fig. 3.3).

The STRO-1+cell fraction can be used directly for *in vitro* assays. Alternatively the STRO-1 population can be further enriched by two-colour FACS as described above (p. 33). This is achieved by incubating the cells with an IgG Mab specific to any cell surface marker of interest for 45 minutes on ice. The cells are then washed twice with HBSS+5% FCS and incubated with a 1/50 dilution of the PE-conjugated secondary antibody (goat anti-mouse IgG; γ-chain specific) for 20 minutes on ice. The cells are then washed a further two times in HBSS+5% FCS prior to further analysis or sorting by flow cytometry.

Studies of BM CFU-F and their progeny in culture *in vitro*

Fibroblast colony-forming cell (CFU-F) assay

The CFU-F assay provides a means to quantify stromal precursor cells *in vitro*. The assay was originally adapted for human studies by Castro-Malaspina *et al.* (1980).

Materials

Culture dishes (Costar, Cambridge, MA)
Alpha modification of Eagle's Medium (α-MEM) (Flow Laboratories, Irvine, Scotland)
FCS 20% batch-tested
L-glutamine 2 mM
β-mercaptoethanol 5×10^{-5} M
Penicillin 100 U/ml and streptomycin 100 μg/ml
RPMI-1640
Paraformaldehyde 1% in PBS
Toluidine blue 0.1% in 1% paraformaldehyde
Olympus SZ-PT dissecting microscope and an IMT-2 inverted light microscope (Olympus Optical Company Co. Ltd, Tokyo, Japan)
Insulin 10 μg/ml bovine pancreas derived (Sigma, St Louis, MO)
Low density lipoprotein (LDL) 80 μg/ml (Sigma, L121139)
Iron saturated human transferrin 200 μg/ml (Sigma)
Dexamethasone sodium phosphate (DEX) 10^{-8}M (David Bull Laboratory, Sydney, Australia)
L-ascorbic acid 2-phosphate (ASC-2P) (Novachem, Melbourne, Australia)
Epidermal growth factor (EGF) purified mouse (Sigma, St Louis, MO)
Platelet derived growth factor (PDGF-BB Pepro Tech Inc., Rocky Hill, NJ)
Fibronectin human (Boehringer Mannheim)

Serum replete culture conditions

BMMNC ($5 \times 10^4 - 50 \times 10^4$ cells/ml) are plated in 24-well tissue culture dishes in α-MEM supplemented with 20% FCS, 2 mM L-glutamine, 5×10^{-5} M β-mercaptoethanol, 100 U/ml penicillin and 100 μg/ml streptomycin. The FCS used must be batch tested for optimal cloning efficiency and colony growth. Cultures are established in triplicate and incubated at 37 °C in a humidified incubator in 5% CO_2 for 14 days. Day 14 cultures are washed twice with warm RPMI-1640 and then fixed for 20 minutes in 1% paraformaldehyde in PBS. Once fixed, the cultures are stained with 0.1% toluidine blue (in 1% paraformaldehyde) for 1 hour and then rinsed in tap water. Aggregates of 50 cells or more are scored as a CFU-F using an Olympus SZ-PT dissecting light microscope.

Serum deprived (SDM) culture conditions

A serum deprived medium (SDM) has recently been developed in which clonogenic growth of CFU-F is absolutely dependent on an exogenous source of growth factors (Gronthos & Simmons, 1995).

BMMNC ($5 \times 10^4 - 50 \times 10^4$ cells/ml) are plated on to 24-well tissue culture dishes in α-MEM supplemented with 10 μg/ml bovine pancreas derived insulin, 2% BSA, 80 μg/ml human LDL, 200 μg/ml iron saturated human transferrin, 2 mM L-glutamine,10^{-8} M DEX, 100 μM ASC-2P, 5×10^{-5} M β-mercaptoethanol, 100 U/ml penicillin and 100 μg/ml streptomycin. To promote attachment of CFU-F under SDM conditions, the dishes are pre-coated with a solution of purified human fibronectin in PBS (10 μg/cm²) for 2 hours at room temperature. Immediately prior to plating the cells, excess fibronectin is removed by washing once with PBS. Growth of human BM CFU-F in this system was found to require the addition of 10 ng/ml of purified mouse EGF and recombinant human PDGF-BB (Gronthos & Simmons, 1995). In other respects, cultures established in SDM are maintained as described above for serum replete cultures. After 14 days, the cultures are fixed, and CFU-F colonies stained and scored according to the previously described criteria.

Clonogenic growth of CFU-F in SDM is comparable to, or better than, that observed in 20% FCS (Fig. 3.1). The development of a reproducible and stringent culture system for the growth and assay of stromal precursors under serum deprived conditions represents an important prerequisite for future studies of the role of growth factors in the regulation of stromal cell proliferation, differentiation and development.

In situ staining of cultured stromal cells derived from BM SPC

The fibroblast-like cells derived from human STRO-1+BMMNC express a variety of cell surface markers including, CD10, CD13, Thy-1, VCAM-1, CD29, CD49b/CD29, CD49c/CD29, CD49e/CD29, CD29 and receptors for PDGF, EGF and IGF-1 (Gronthos & Simmons, 1995, and unpublished observations). In other studies it has been shown that the progeny of cultured human CFU-F express a wide variety of integrins, including CD29, CD49a/CD29, CD49b/CD29, CD49c/CD29 and CD51 (Soligo *et al.*, 1990) and produce collagen types I, III, IV and V, laminin and fibronectin (Castro-Malaspina *et al.*, 1980; Lim *et al.*, 1986; Perkins & Fleischman, 1990). All of these proteins are widely expressed, however, and the only marker thus far which appears to be restricted to SPC is the STRO-1 antigen.

There are clear differences between the cell surface phenotype of SPC *in vivo* and CFU-F and their progeny *in vitro*, an example of the problems faced when using cell culture models to mimic *in vivo* conditions. The integrin molecule CD49c/CD29 is expressed *in vitro* but not *in vivo*. Although STRO-1 is highly expressed on CFU-F its expression rapidly disappears during culture *in vitro* from all but a minor population of cells. The significance of this is uncertain, although it may be related to increased differentiation since STRO-1 expression is not detectable in more mature stromal cell types (Simmons & Tork-Storb, 1991*a*; Gronthos *et al.*, 1994).

The method used for the detection of cell surface and cytoplasmic antigens *in situ* using immunoperoxidase staining is described here.

Materials

Paraformaldehyde 2% in PBS
Methanol
Vectorstain ABC kits, PK-4009 and PK-4010, and Peroxidase substrate kit
    AEC, SK-4200 (Vector Laboratories)
Olympus IMT-2 inverted light microscope

Cultures of stromal cells derived from STRO-1+BMMNC are washed three times in PBS and then fixed in 2% paraformaldehyde for 20 minutes at room temperature. They are then washed three times in PBS before being permeabilised in methanol for 15 minutes at −20 °C. The fixed and permeabilised cells are then washed three times in PBS and incubated with the STRO-1 antibody for 1 hour at room temperature. Non-specific staining is assessed by treating control wells under identical conditions with 1A6.12, an isotype-matched monoclonal antibody of irrelevant specificity. The cells are then washed three times in PBS and specifically bound antibody localised using a Vectorstain ABC kit for mouse IgM (PK-4009) and a peroxidase substrate kit AEC (SK-4200) according to the manufacturer's instructions. The same protocol can be used for IgG mouse Mabs by substituting the second goat anti-mouse IgM specific biotinylated antibody for a goat anti-mouse IgG specific antibody conjugated to biotin using Vectorstain ABC kit (PK-4010). Following immunostaining, the cells are counter stained with Mayer's haematoxylin and examined using an Olympus IMT-2 inverted light microscope.

Osteogenic differentiation of human BM CFU-F

Materials

α-MEM (Flow Laboratories Irvine Scotland) and supplements as stated
KH_2PO_4 (BDH Chemicals)
T-75 culture flasks

It has recently been demonstrated that human STRO-1+BM CFU-F have the capacity to differentiate into functional osteoblast-like cells (Gronthos *et al.*, 1994). Osteogenesis was assessed by: (a) the expression of alkaline phosphatase activity, (b) the expression of osteocalcin (mRNA and protein) and (c) the development of a mineralised, hydroxyapatite-containing extracellular matrix.

STRO-1+BMMNC (2×10^4/flask) are cultured in α-MEM supplemented with 20% FCS, 2 mM L-glutamine, 5×10^{-5} M β-mercaptoethanol, 100 μM ASC-2P, 100 U/ml penicillin, and 100 μg/ml streptomycin at 37 °C, 5% CO_2 for 7 days. The culture medium is then replaced with α-MEM supplemented as described above, but with 10% FCS and without β-mercaptoethanol, and with the addition of 10 mM HEPES, 10^{-8} M DEX and 1.8 mM KH_2PO_4 to give a final phosphate concentration of 2.9 mM. Thereafter, the culture medium is replaced twice weekly for 4–6 weeks in order to induce matrix mineralisation. In addition to the formation of a mineralised matrix, the differentiation of adipocytes has also been observed under these conditions of culture.

Cryopreservation of BM CFU-F

Single cell suspensions of cultured stromal cells are prepared by trypsin digest as described previously (p. 35). The trypsinised cells are then diluted and washed in cold HBSS+5% FCS. Following centrifugation, the cell pellet is resuspended at a concentration of 2×10^7 cells per ml in FCS and maintained on ice. An equal volume of freeze mix (20% DMSO in cold FCS) is then added gradually while gently mixing the cells to give a final concentration 1×10^7 cells/ml in 10% DMSO. The cells are aliquoted into 1.8 ml cryovials (NUNC; 1 ml/tube) on ice and then frozen at a rate of -1 °C per minute using a rate control freezer. The frozen vials are transferred to liquid nitrogen for long-term storage. Recovery of the frozen stocks is achieved by rapidly thawing the cells in a 37 °C water bath. They are then resuspended and washed in cold HBSS+5% FCS. To assess viability, a small aliquot of the recovered cells is diluted 1:5 in 0.4% trypan blue in PBS and the number of viable cells determined using a haemocytometer. Typically, this procedure gives viabilities of between 60 and 70%.

References

Allen, T.D., Dexter, T.M. & Simmons, P.J. (1990). Marrow biology and stem cells. *Immunol. Ser.*, **49**, 1–38.

Bennett, J.H., Joyner, C.J., Triffitt, J.T. & Owen, M.E. (1991). Adipocyte cells cultured from marrow have osteogenic potential. *J. Cell Sci.*, **99**, 131-9.

Castro-Malaspina, H., Gay, R.E., Resnick, G., Kapoor, N., Meyers, P., Chiarieri, D., McKenzie, S., Broxmeyer, H.E. & Moore, M.A.S. (1980). Characterization of human bone marrow fibroblast colony-forming cells (CFU-F) and their progeny. *Blood*, **56**, 289-301.

Dexter, T.M., Spooncer, E., Simmons, P.J. & Allen, T.D. (1984). Long-term marrow culture: an overview of techniques and experience. In *Kroc Foundation Series*, Vol 18, pp. 57–96. New York: Alan R. Liss, Inc.

Friedenstein, A.J. (1980). Stromal mechanisms of bone marrow: cloning *in vitro* and retransplantation *in vivo*. In *Immunology of Bone Marrow Transplantation*, ed. S. Thienfelder, pp. 19–29. Berlin: Springer-Verlag.

Friedenstein, A.J., Chailakhyan, R.K. & Lalykina, K.S. (1970). The development of fibroblast colonies in monolayer cultures of guinea pig bone marrow and spleen cells. *Cell Tissue Kinet.*, **3**, 393-403.

Friedenstein, A.J., Chailakhyan, R.K. & Gerasimov, U.V. (1987). Bone marrow osteogenic stem cells: *in vitro* cultivation and transplantation in diffusion chambers. *Cell Tissue Kinet.*, **20**, 263–72.

Gronthos, S. & Simmons, P.J. (1995). The growth factor requirements of STRO-1+human bone marrow stromal precursors under serum-deprived conditions. *Blood*, **85**, 929-40.

Gronthos, S. & Simmons, P.J. (1996). The biology and application of human bone marrow stromal cell precursor. *J. Hematother.*, **5**, 15-23.

Gronthos, S., Ohta, S., Graves, S.E. & Simmons, P.J. (1994). The STRO-1+fraction of adult human bone marrow contains the osteogenic precursors. *Blood*, **84**, 4164–73.

Huang, S. & Terstappen, L.W.M.M. (1992). Formation of haematopoietic microenvironment and haematopoietic stem cells from single human bone marrow stem cells. *Nature*, **360**, 745–9.

Lichtman, M.A. (1981). The ultrastructure of the hemopoietic environment of the marrow: a review. *Exp. Hemat.*, **9**, 391–410.

Lim, B., Izaguirre, C.A., Aye M.T., Huebsch, L., Drouin, J., Richardson, C., Minden, M.D. & Messner, H.A. (1986). Characterization of reticulofibroblastoid colonies (CFU-RF) derived from bone marrow and long-term marrow culture monolayers. *J. Cell Physiol.*, **127**, 45–54.

Miltenyi, S., Muller, W., Weichel, W. & Radbruch, A. (1990). High gradient magnetic cell separation with MACS. *Cytometry*, **11**, 231–8.

Patt, H.M. & Maloney, M.A. (1975). Bone marrow regeneration after local injury: a review. *Exp. Hemat.*, **3**, 135–48.

Perkins, S. & Fleischman, R.A. (1990). Stromal cell progeny of murine bone marrow fibroblast colony-forming units are clonal endothelial-like cells that express collagen IV and laminin. *Blood*, **75**, 620–5.

Simmons, P.J. & Torok-Storb, B. (1991*a*). Identification of stromal cell precursors in human bone marrow by a novel monoclonal antibody, STRO-1. *Blood*, **78**, 55–62.

Simmons, P.J. & Torok-Storb, B. (1991*b*). CD34 expression by stromal precursors in normal human adult bone marrow. *Blood*, **78**, 2848–53.

Simmons, P.J., Przepiorka, D., Donnall Thomas, E. & Torok-Storb, B. (1987). Host origin of marrow stromal cells following allogeneic bone marrow transplantation, *Nature*, **328**, 429–32.

Simmons, P.J., Gronthos, S., Zannettino, A., Ohta, S. & Graves, S.E (1994). Isolation, characterization and functional activity of human marrow stromal progenitors in hemopoiesis. *Progr. Clin. Biol. Res.*, **389**, 271–80.

Soligo, D., Schiro, R., Luksch, R., Manara, G. & Quirici, N. (1990). Expression of integrins in human bone marrow. *Br. J. Haemat.*, **76**, 323–32.

Tavassoli, M. & Friedenstein, A. (1983). Hemopoietic stromal microenvironment. *Ann. J. Hemat.*, **15**, 195–203.

Testa, N.G., Hendry, J.H. & Molineux, G. (1988). Long-term bone marrow damage after cytotoxic treatment: stem cells and microenvironment. In *Hematopoiesis: Long-term Effects of Chemotherapy and Radiation*, ed. N.G. Testa & R.P. Gale, Hematology, vol. 8, pp. 75–92. New York and Basel: Marcel Dekker, Inc.

Van den Heuvel, R.L., Versele, S.R.M., Schoeters, J.E.R. & Vanderborght, O.L.J. (1987). Stromal stem cells (CFU-F) in yolk sac, liver, spleen and bone marrow of pre- and postnatal mice. *Br. J. Haemat.*, **66**, 15–20.

Waller, E.K., Olweus, J., Lund-Johansen, F., Huang, S., Nguyen, M., Guo, G-R. & Terstappen, L.W.M.M. (1995). The 'common stem cell' hypothesis reevaluated: human fetal bone marrow contains separate populations of hematopoietic and stromal progenitors. *Blood*, **85**, 2422–35.

4

Isolation and culture of human bone-derived cells

Roger Gundle, Karina Stewart, Joanne Screen
and Jon N. Beresford

Introduction

Cells of the osteoblast lineage are essential for the normal growth, development and maintenance of the vertebrate skeleton. They are responsible for bone formation, participate in the regulation of bone resorption and contribute to the maintenance of the haematopoietic microenvironment. The term 'osteoblast' refers to a relatively short-lived cell that is found *in vivo* in direct apposition to a newly forming bone surface and one which is actively engaged in the synthesis and secretion of a collagen-rich extracellular matrix (osteoid). In histology texts osteoblasts are frequently described as plump, basophilic cells that are obviously polarised and, as befits their function, possess a well-developed golgi apparatus and endoplasmic reticulum.

It is important to appreciate that osteoblasts are not terminally differentiated cells. During the process of bone formation, a proportion of osteoblasts become entrapped within the bone matrix and develop long, cellular processes that enable them to make contact with other cells; these are referred to as osteocytes and they are the most abundant cell type present in adult human bone. Of the remaining cells, the majority are destined to die, by a mechanism that is morphologically inconspicuous and therefore bears the hallmark of apoptosis. A minority of osteoblasts cease their biosynthetic activities, develop a flattened, elongated morphology and persist as bone lining cells (Parfitt, 1990).

Much of our knowledge of the biology of cells of the osteoblast lineage is based on studies using fetal or neonatal tissue from experimental animals, which is poorly mineralised and highly cellular when compared with adult bone. Many of the clinically important bone diseases (for example, osteoporosis and Paget's disease), affect the adult skeleton, and the majority do not normally occur in the animal population. Thus, there are clear advantages in studying cells isolated from adult human bone.

The development of techniques for the isolation and culture of adult human cells of the osteoblast lineage is a relatively recent phenomenon. Although sporadic examples can be found in the literature prior to 1980 (Bard *et al.*, 1974; Mills *et al.*, 1979), it was not until the middle 1980s that several groups of investigators embarked upon a systematic evaluation of the phenotypic characteristics of cultured human bone-derived cells (HBDC) and the conditions of culture that favoured their proliferation and osteogenic differentiation (for review see Gallagher, Gundle & Beresford, 1996).

The purpose of this chapter is to describe the methodology currently in use in the authors' laboratories for the isolation and culture of cells of the osteoblast lineage from explants of adult human trabecular bone, a technique developed originally by Beresford *et al.* (Gallagher *et al.*, 1983; Beresford *et al.*, 1984; Beresford, Gallagher & Russell, 1986). A detailed description of alternative methods that have been developed (Wergedal & Baylink, 1984; Gehron Robey & Termine, 1985; Marie *et al.*, 1989), can be found in Gallagher, Gundle and Beresford, 1996.

Potential clinical applications

Fracture non-union, prosthetic loosening with bone loss, and the replacement of large bony defects, are all common problems in clinical orthopaedics. The traditional approach to dealing with these has involved the use of autografts or allografts, neither of which is entirely satisfactory. An alternative approach to the resolution of these problems is the use of culture-expanded autologous bone-derived cells in combination with biocompatible implant materials (for review, see Bruder, Fink & Caplan, 1994). Although limited trials have already begun (Osipian *et al.*, 1987), it is undoubtedly the case that the more widespread use of autologous bone-derived cells will depend on continued improvements in the methodology employed for their isolation and culture and on our understanding of the factors that regulate their proliferation and osteogenic differentiation.

Characterisation of human bone-derived cells (HBDC)

Morphology, proliferative potential and differentiated characteristics

The nomenclature used for HBDC from bone explant cultures is as follows. The first flask of cells to grow from fresh explants is referred to as E1 cells (primary culture); after passage of these cells, they are denoted as E1P1 (secondary culture). If fresh medium is put on the explants after cell passage, or if the explants are transferred into a new flask at the end of primary culture, then a further flask of cells can be grown and these are known as E2 cells, and, after passage, as E2P1.

HBDC at E1P1 have been characterised extensively and shown to possess a phenotype that is clearly distinct from that of fibroblastic cells derived from the dermis of skin (HSDC) obtained from the same donors (Gallagher, Gundle & Beresford, 1996). In cultures of HBDC, the predominant cell type has a flattened, multipolar morphology and an abundance of cytoplasmic stress fibres. In contrast, in cultures of HSDC, the predominant cell type has an elongated, bipolar morphology. These morphological differences are apparent in pre- as well as post-confluent cultures (Fig. 4.1, a–d). When cultured in the presence of a physiological level of glucocorticoid (pp. 50–53), HBDCs assume a more polygonal morphology. This shape change has been interpreted by some investigators as indicating that the cells have developed a more mature, 'osteoblast-like' phenotype (Wong et al., 1990).

When compared with HSDC obtained from the same donors, HBDC proliferate less rapidly, reach lower saturation densities, and express high levels of the tissue non-specific isoform of the enzyme alkaline phosphatase (AP). Both cell types synthesise a collagen-rich extracellular matrix, but that produced by HBDC is composed predominantly of type I collagen with typically $\leq$10% collagen type III, the bulk of which is secreted into the culture medium (Fig. 4.2). HBDC also synthesise small quantities of collagen type V, the presence of which has been confirmed recently by cyanogen bromide peptide mapping.

In addition to collagens, HBDC synthesise several non-collagenous proteins that are present in the extracellular bone matrix. These include bone sialoprotein (BSP) and osteocalcin (OC), both of which are specific, late stage markers for cells of the osteoblast lineage. The level of expression of these proteins, as well as that of AP, is influenced markedly by the cells' state of maturation, which is a function of cell density, and by the absence, or presence, of specific hormones, in particular, glucocorticoids and $1,25(OH)_2D_3$, the active metabolite of vitamin D_3 (Beresford, Graves & Smoothy, 1993; Gallagher, Gundle & Beresford, 1996).

The most compelling evidence for the presence of cells of the osteogenic lineage in cultures of HBDC is the demonstration of matrix mineralisation in vitro and the formation of a well-organised, mineralised tissue that histologically resembles bone when these same cells are implanted in vivo within diffusion chambers (Gundle & Beresford, 1995; Gundle, Joyner & Triffitt, 1995).

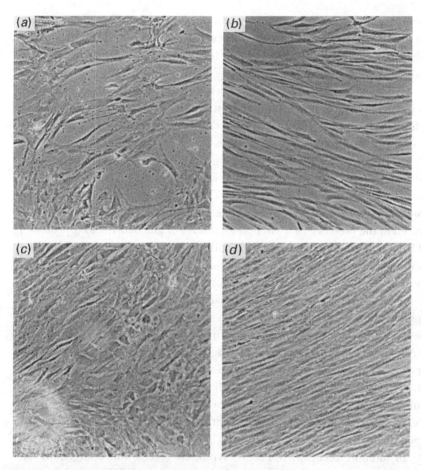

Fig. 4.1. Morphological differences in culture between human bone- and skin-derived cells obtained from the same donor. (*a*), (*b*), subconfluent cultures; (*c*), (*d*) confluent cultures. (*a*), (*c*), HBDC; (*b*), (*d*), HSDC. Original magnification ×100.

Factors influencing the proliferation and differentiation of human bone-derived cells

Ascorbic acid (Vitamin C)

Cells of the osteoblast lineage secrete an extracellular matrix that is rich in type I collagen, and factors which regulate the synthesis, secretion and extracellular processing of procollagen have been shown to influence profoundly the proliferation and differentiation of osteoblast precursors in a variety of animal bone-derived cell culture systems (Beresford *et al.*, 1993). L-Ascorbic

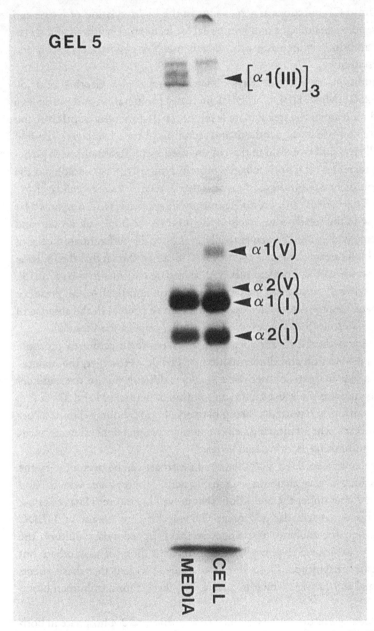

Fig. 4.2. Collagen types synthesised by cultured human bone-derived cells. SDS-PAGE of purified radiolabelled collagens.

acid (vitamin C), functions as a co-factor in the hydroxylation of lysine and proline residues in collagen and is essential for its normal synthesis and secretion. In addition, it increases procollagen mRNA gene transcription and mRNA stability.

The addition on alternate days of freshly prepared L-ascorbic acid (50 μg/ml=250 μM) to HBDC in E1P1 increases proliferation and produces a sustained increase in the steady state levels of α1 (I)-procollagen mRNA and a dramatic increase in the synthesis and secretion of type I collagen (six–fold in the cell layer and five–fold in the culture medium). Treatment with ascorbate also increases total non-collagen protein synthesis ($\sim$ two–fold) and the proportion of newly synthesised protein that is retained in the cell layer (67 vs 42%). The increase in non-collagen protein synthesis is accompanied by an increase in the steady-state levels of the mRNA for bone sialoprotein and osteocalcin, which are characteristically expressed by differentiated cells of the osteoblast lineage. Using a specific RIA, it can be shown that the increase in the expression of osteocalcin mRNA, following treatment with calcitriol, is associated with an increase in protein secretion and that, in the presence of ascorbate, there is detectable basal production of protein in the absence of added seco-steroid (S.E. Graves and R. Gundle, unpublished results).

The changes described are consistent with ascorbate acting to promote both the proliferation and differentiation of HBDC. However, the maintenance of adequate levels of ascorbate *in vitro* is difficult due to the naturally occurring vitamins' excessive lability in solution at neutral pH and 37 °C ($t_{1/2}$ in culture variously reported as being between 1 and 6 hours). This is a particular problem when culturing cells of human origin, since they are incapable of synthesising ascorbic acid *de novo*.

The problems associated with the use of ascorbate can be overcome by the use of its long-acting analogue L-ascorbic acid 2–phosphate, which has a half-life of 7 days under the conditions that prevail in culture (Hata & Senoo, 1989). This is a particular advantage during the early stages of HBDC culture, when the medium is changed weekly. In secondary culture, the effects of ascorbic acid 2–phosphate are similar to those of L-ascorbate but of far greater magnitude. Dose–response studies revealed that these effects are maximal at 100 μM, which is close to the levels found in human blood (34–68 μM).

The effects of culture in the presence of ascorbic acid 2–phosphate in both primary and secondary culture are even more dramatic and include an $\sim$ nine-fold increase in cell number (three-fold in primary and three-fold in secondary) and a further increase in the synthesis of collagenous and non-collagenous protein, the bulk of which is retained in the cell layer (summar-

Table 4.1. *The influence of ascorbate on protein synthesis and the expression of the osteoblast phenotype in cultures of HBDC*

Parameter	Treated/control ratio[a]
CDP (dpm/μg DNA)[c]	
Cell layer	2.2
Culture medium	1.2
NCP (dpm/μg DNA)	
Cell layer	2.2
Culture medium	1.1
Alkaline phosphatase (units/μg DNA)	0.3
α1(I)-procollagen mRNA	1.0
Bone sialoprotein mRNA	1.0
Osteocalcin mRNA[b]	3.3

Notes:
Collagenous (CDP) and non-collagenous (NCP) protein synthesis and alkaline phosphatase activity were determined at the end of 35 days (28 days in primary culture and 7 days in secondary culture). mRNA expression was determined at the end of 47 days (28 days in primary culture and 19 days in secondary culture). Cells from separate donors were used in each experiment.
[a] Ascorbate in primary and secondary culture/ascorbate in secondary culture only.
[b] In cultures treated with 10 nM $1,25(OH)_2D_3$.
[c] dpm (disintegrations per minute, labelled with tritiated proline).

ised in Table 4.1). The expression of AP, however, is reduced, even when compared with cells cultured in the complete absence of ascorbate (Fig. 4.3). This suggests that the cells in the cultures treated continuously with ascorbate have progressed to a late stage of osteoblast differentiation. Consistent with this possibility, it has been shown that these cultures also express much higher basal and $1,25(OH)_2D_3$-stimulated levels of OC mRNA and protein (S.E. Graves and R. Gundle, unpublished results) (Table 4.1).

An observation of particular importance is that the addition of ascorbate in secondary culture, even for periods of up to 28 days, cannot compensate fully for its omission in primary culture. This suggests that maintaining adequate levels of ascorbate, and hence matrix synthesis, during the early stages of explant culture is of critical importance for the survival of cells that retain the ability to proliferate extensively and give rise to precursors capable of undergoing osteogenic differentiation.

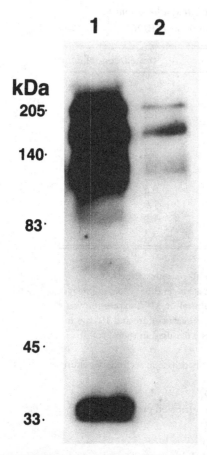

Fig. 4.3. The influence of ascorbate on the expression of alkaline phosphatase in cultures of human bone-derived cells. Lysates were prepared from cells cultured in the absence (lane 1) or presence (lane 2) of ascorbate, and equivalent amounts of protein from each separated by SDS-PAGE. Following transfer to nitrocellulose, AP was detected using the monoclonal antibody B4-78 in conjunction with a peroxidase-labelled secondary antibody and enhanced chemiluminescence.

Glucocorticoids

In many bone-derived cell culture systems, osteogenic differentiation is glucocorticoid dependent. The degree of glucocorticoid dependence varies according to the state of maturation of the cells, which reflects, at least in part, their origin (bone surfaces vs bone marrow) and the species under investigation (Aubin & Herbertson, Chapter 6).

Table 4.2. *The influence of glucocorticoids on proliferation, protein synthesis and the activity of alkaline phosphatase in cultures of HBDC*

Parameter	Treated/control ratio
Proliferation (µg DNA/well)	1.1 ± 0.3
Total CDP (dpm/µg DNA)	1.2 ± 0.3
Total NCP (dpm/µg DNA)	1.4 ± 0.2
Alkaline phosphatase (units/µg DNA)	10 ± 2.0

Note:
DNA, collagenous (CDP) and non-collagenous (NCP) protein synthesis and alkaline phosphatase activity were determined at the end of 43 days (35 days in primary culture and 8 days in secondary culture). The data are presented as the mean T/C ratio±SD for n=three donors for cells cultured in the presence (T) and absence (C) of 10 nM dexamethasone.

Early studies using HBDC revealed that long-term culture in the presence of glucocorticoids, whilst enhancing some indices of osteoblast maturation, decreased collagen synthesis and inhibited cell proliferation (Beresford *et al.*, 1993; Wong *et al.*, 1990). However, when HBDC are cultured in the presence of the long-acting ascorbate analogue, these inhibitory effects of glucocorticoids are negated and, in addition to expressing high levels of AP activity, the cells secrete a dense extracellular matrix, which will mineralise in the presence of added phosphate (Table 4.2) (Gundle, 1995; Gundle & Beresford, 1995; Gallagher *et al.*, 1996).

Using cells derived from a large number of donors it has been established that optimal proliferation and differentiation of HBDC occurs in the presence of 10 nM dexamethasone or 200 nM hydrocortisone (Sigma Aldrich Chemical Co. Ltd., Cat. No. H 2270 and D 8893, respectively), both of which approximate to a physiological level of glucocorticoid.

In addition to promoting the expression of the differentiated phenotype by cells of the osteoblast lineage, glucocorticoids also appear able to promote the proliferation and/or survival of primitive cells expressing the STRO-1 antigen in cultures of HBDC (Gronthos, Graves & Simmons, Chapter 3). When the cells harvested from control and glucocorticoid-treated cultures are labelled with monoclonal antibodies recognising the STRO-1 antigen and AP and then analysed by flow cytometry, four subpopulations of cells are detected: STRO-1-/AP-, STRO-1+/AP-, STRO-1+/AP+ and STRO-1-/AP+ (Fig. 4.4). In control cultures, cells in the double negative fraction predominate, with STRO-1+/AP- cells comprising typically

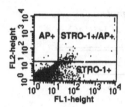

Isotype-matched control antibodies

Quad	Events	% total	X Geomean	Y Geomean
UL	5	0.10	8.54	21.87
UR	102	2.04	73.79	40.15
LL	4705	94.10	3.10	2.18
LR	188	3.76	42.96	3.41

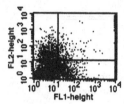

Control

Quad	Events	% total	X Geomean	Y Geomean
UL	636	12.72	4.92	44.95
UR	216	4.32	59.71	62.93
LL	3573	71.46	3.94	2.69
LR	575	11.50	40.78	3.38

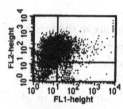

10 nM Dexamethasone

Quad	Events	% total	X Geomean	Y Geomean
UL	2625	52.50	6.46	74.97
UR	1487	29.74	36.76	103.00
LL	695	13.90	3.73	4.72
LR	193	3.86	47.29	4.30

Fig. 4.4. Co-expression of the STRO-1 antigen and AP in primary cultures of human bone-derived cells. HBDC cultured for 28 days in the absence or presence of 10 nM dexamethasone were dual-labelled with the monoclonal antibodies STRO-1 and B4–78 and analysed by flow cytometry. STRO-1 binding is plotted on the abscissa (FL1) and that of B4-78 on the ordinate (FL2). UL, UR, LL and LR refer to the proportion of cells present in the gated cell populations (STRO-1-/AP+, STRO-1+/AP+, STRO-1-/AP−and STRO-1+/AP−, respectively).

≤15% of total. Continuous culture in the presence of glucocorticoid is associated consistently with an increase in the proportion of cells expressing the STRO-1 antigen, the majority of which co-express AP (Fig. 4.4).

It is proposed that the subpopulations identified by dual labelling with the anti-STRO-1 and AP monoclonal antibodies form part of a lineage progression. Cells in the STRO-1+/AP- subpopulation are presumed to represent multipotential precursors and those in the STRO-1+/AP+subpopulation committed osteogenic precursors. The dramatic increase in the size of this dual-labelled population in glucocorticoid-treated cultures is consistent with this possibility. Cells in the STRO-1-/AP+subpopulation are presumed to be maturing osteoblasts. That the cells progress from being STRO-1+/APto STRO-1+/AP+ and then to STRO-1-/AP+ is supported by the results of studies in which the proportion of cells present in the different subpopulations was determined at different times following the addition of glucocorticoid (Stewart *et al.*, 1996). These showed clearly that the initial event following treatment with glucocorticoid is an increase in the size of the STRO-1+/AP- subpopulation of cells. With increasing duration of treatment, the size of this subpopulation declines concomitant with an increase in that of the STRO-1+/AP+subpopulation. At the later time points still, the size of this subpopulation also declines and this is associated with a corresponding increase in the size of the STRO-1-/AP+subpopulation.

By combining FACS with RT-PCR, it has been possible to investigate the expression of osteoblast-related mRNAs in the different cell subpopulations. The expression of OC and BSP, both late stage markers of the osteoblast phenotype, was highest in the STRO-1-/AP+subpopulation, as was that of the oestrogen receptor mRNA, irrespective of the presence or absence of glucocorticoid. In the absence of glucocorticoid, parathyroid hormone receptor mRNA was expressed only in the STRO-1+/AP+ & STRO-1-/AP+subpopulations (Stewart, *et al.*, 1996; K. Stewart, J. Screen & J.N. Beresford, unpublished observations). These results, although yet to be confirmed at the level of protein synthesis, are entirely consistent with the possibility that the cells in the different subpopulations represent cells of the osteoblast lineage at different stages of differentiation.

The least well-defined subpopulation in the HBDC cultures are the cells which express neither STRO-1 nor AP on their surfaces. The results of preliminary investigations suggest, however, that this subpopulation may contain very primitive cells as well as terminally differentiated (OC-positive /AP-negative) cells of the osteoblast lineage (Stewart *et al.*, 1996).

Isolation and culture of human bone derived cells (HBDC)

Establishment of primary cultures

Materials

Dulbecco's phosphate buffered saline, pH 7.4 (PBS) (GIBCO/BRL, Cat. No. 14200-059)
Bone curette
Stainless steel scalpel blade with integral handle (Merck, Cat. No. 406/0024/00)
Fine scissors

50 ml polypropylene tube (Becton Dickinson 'Blue Max', Cat. No. 2098)

80 cm^2 flask (Becton Dickinson, Cat. No. 3084 or 3110 or Nunc, available from GIBCO/BRL Life Technologies Ltd., Cat. No. 1-53732A)

Heat inactivated fetal calf serum (FCS)

25–50 U/ml penicillin and 25–50 μg/ml streptomycin

50 μg/ml L-ascorbic acid or 100 μM L-ascorbic acid 2-phosphate (Alpha Laboratories Ltd., Cat. No. 013-12061)

HEPES buffered Dulbecco's modification of EMEM with 4500 g/litre glucose (10× liquid concentrate, Gibco/BRL, Cat. No. 12501-029 or dry powder formulation, ICN, Cat. No. 10-331-20)

Sodium bicarbonate, HEPES (free acid), L-glutamine (Sigma Aldrich Chemical Co. Ltd, Cat. Nos. S 5761, H O763 and G 5763, respectively)

Tissue removed at surgery or biopsy is transferred to a sterile container filled with PBS or serum-free medium (SFM) for transport to the laboratory. The specimen should be collected and processed with minimal delay, preferably the same day. Storage of the specimen for periods of up to 24 h at 4 °C in PBS or SFM, however, does not noticeably alter the ability of the tissue to give rise to proliferative precursors when cultured *in vitro*.

Ideally, the bone used should be radiologically normal. An excellent source is the upper femur of patients undergoing total hip replacement surgery for osteoarthritis. Trabecular bone is removed from this site prior to the insertion of the femoral prosthesis and would otherwise be discarded. The tissue obtained is remote from the hip joint itself, and thus from the site of pathology, and is free of contaminating soft tissue. If cortical bone is present, remove extraneous soft connective tissue from the outer surfaces of the bone by scraping with a sterile scalpel blade. Rinse the tissue in sterile PBS and transfer to a sterile Petri dish containing a small volume of PBS (5–20 ml depending on the size of the specimen).

To gain access to the cancellous bone segments of long bone or rib, split these longitudinally with the aid of sterile surgical bone cutters. Cancellous bone is then removed using a bone curette. To maximise the recovery of tissue, the endosteal surface of the bone can be scraped with a scalpel blade. Disposable scalpel blades have an alarming tendency to shatter during this process and it is advisable therefore, to use a solid, stainless steel blade with integral handle. For larger specimens, femoral heads or sections of femur, cancellous bone can be removed directly from the open ends using a bone curette.

Transfer the cancellous bone fragments to a clean Petri dish containing 2–3 ml of PBS and dice into pieces 3–5 mm in diameter. This can be achieved in two stages, using first a scalpel blade and then fine scissors. Decant the PBS and transfer the bone chips to a 50 ml polypropylene tube containing 15–20 ml of PBS. Vortex the tube vigorously 3×10 seconds and then leave to stand for 30 seconds to allow the bone fragments to settle. Carefully decant the supernatant containing haematopoietic tissue and dislodged cells, add a further 15–20 ml of PBS and vortex the bone fragments as before. Repeat this process a minimum of three times, or until no remaining haematopoietic marrow is visible and the bone fragments have assumed a white, ivory-like appearance.

Culture the washed bone fragments as explants (0.2–0.6 g of tissue per 80 cm^2 flask) at 37 °C in an humidified atmosphere of 95% air 5% CO_2 in 10 ml medium (see below) containing 10% (v/v) heat inactivated fetal calf serum, 25–50 U/ml penicillin, 25–50 µg/ml streptomycin and 50 µg/ml freshly prepared L-ascorbic acid or 100 µM L-ascorbic acid 2–phosphate. The inclusion of ascorbate as a standard supplement is a recent and important modification to the original cell culture protocol. The rationale for this is discussed on pp. 46–50. Leave the cultures undisturbed for 5–7 days, after which time the medium is replaced with an equal volume of fresh medium taking care not to dislodge the explants. Cultures are fed again at 14 days and thereafter twice or thrice weekly.

With the exception of small numbers of isolated cells, which probably become detached from the bone surface during the dissection, the first evidence of cellular proliferation is observed on the surface of the explants and this normally occurs within 5–7 days of plating. Between 7 and 10 days cells can be observed migrating from the explants on to the surface of the culture dish. If care is taken not to dislodge the explants when feeding, and they are left undisturbed between media changes, they rapidly become anchored to the substratum by the cellular outgrowths. Cultures generally attain confluence 4–6 weeks post-plating.

The preferred medium for culture of HBDC is HEPES buffered Dulbecco's modification of EMEM with 4500 g/l glucose. The latter is prepared according to the manufacturers' instructions, using pyrogen-free water with the addition of the following tissue culture grade reagents: 0.85 g/l of sodium bicarbonate, 4.766 g/l of HEPES (free acid) and 0.29 g/l of L-glutamine.

Batches of serum vary widely in their ability to support the proliferation and differentiation (basal alkaline phosphatase activity) of human bone-derived cells and batches of serum from several suppliers are screened routinely before ordering. As this is a tedious and time-consuming process, it is advisable to reserve a large quantity of serum once a suitable batch has been identified.

Heat inactivation of serum

The serum supplement is heat-inactivated prior to use. Most suppliers will perform this task for an additional fee. Alternatively, 500 ml batches of thawed serum can be heat-inactivated as follows.

1 Place the thawed bottle into a water bath at 37 °C and allow to equilibrate.
2 Increase the temperature to 56 °C and incubate for a further 45 minutes with periodic mixing.
3 Allow the serum to cool and then store frozen at ≤ -20 °C in 50 ml aliquots.

Origin of the initiating cell type

The precise origin of the initiating cell type in cultures of HBDC has not been established. Given the methodology used in the preparation of the explants, however, it is not unreasonable to assume that the cultures are derived from proliferative precursors that are located close to, or in contact with, bone surfaces. Two observations are consistent with this. First, histological examination of the explants post-preparation but prior to culture has revealed that the bone surfaces remain covered with a thin (1–2 cell diameters) layer of cells and matrix with only occasional islands of marrow tissue in the inter-trabecular spaces (Gundle, 1995). Secondly, by histology and phase contrast microscopy, it is evident that cell proliferation first occurs on the surfaces of the explants and that this remains the principal site of cellular activity during the initial stages of culture (≤ 10 days post-explantation).

The identity of the cells present on the surfaces of the explants is unknown, and it is not clear whether they all possess proliferative potential. Recent observations suggest, however, that a proportion of them may be closely related to the clonogenic, multipotential precursors present in the bone marrow. This conclusion is based on the detection of a subpopulation of cells in primary cultures of HBDC (typically $\leq 15\%$; pp. 51–53) that express the STRO-1 antigen (Fig. 4.4). That these cells are capable of giving rise to osteogenic precursors is now well established (Gronthos *et al.*, Chapter 3). Their presence is also consistent with the differentiation of other marrow stromal cell types, principally adipocytes, when HBDC are cultured for extended periods in the presence of glucocorticoids (pp. 57–58).

Presence of other cell types

The presence of other cell types, including endothelial cells and those derived from the haematopoietic stem cell, has been investigated using a large panel of monoclonal antibodies and flow cytometry and/or immunocytochemistry. The results of these studies reveal that, at first passage, there are no detectable endothelial, lymphoid or erythroid cells present. A consistent finding, however, is the presence of small numbers of cells (≤5%) expressing antigens present on cells of the mono-cyte/macrophage series (H. Skojdt & R.G.G. Russell, unpublished observations; S.E. Graves & N.A. Athanasou, unpublished observations).

If primary cultures are maintained for ≥ 6 weeks in the presence of ascorbate and glucocorticoid, the differentiation of adipocytes is observed. Initially, this occurs in the layer of cells and matrix immediately adjacent to the explants. Subsequently, however, the adipocytes can be observed to migrate out from the explants and over the underlying layer of HBDCs (Fig. 4.5). When these late stage cultures are passaged, adipocytes are not immediately apparent, suggesting either that they are lost during this process or that they dedifferentiate and lose their accumulated lipid, but they reappear and rapidly increase in number post-confluence. Studies have shown that the reappearance of adipocytes in secondary culture is prevented when the medium is supplemented with inorganic phosphate (Pi) or β-glycerophosphate (β-GP) to promote matrix mineralisation. This is consistent with the hypothesis that there is an inverse relationship between the differentiation of cells of the adipogenic and osteogenic lineages (Beresford et al., 1992; Bennett et al., 1991) (see also Gimble, Chapter 5).

In post-natal bone marrow, adipocytes and osteoblasts form a part of the stromal system of bone and marrow, all members of which are derived from a common stromal stem cell (Owen, Chapter 1). Thus, the appearance of both cell types in cultures of HBDC lends further support to the hypothesis that these cultures are initiated by a relatively small number of multi-potential, highly proliferative marrow stromal precursors that bear the STRO-1 antigen on their surface (pp. 51–53).

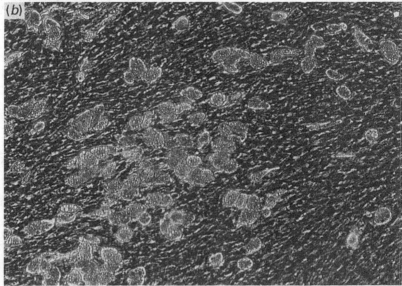

Fig. 4.5. Appearance of adipocytes in a late-stage, primary culture of human bone-
derived cells. The photomicrographs show cells from a primary culture treated
continuously with 100 μM L-ascorbic acid 2-phosphate and 10 nM dexamethasone
for approximately 12 weeks. (*a*), Adipocytes can be seen migrating away from the
surface of an explant of trabecular bone. (*b*), Adipocytes containing numerous and
highly refractile lipid-laden vacuoles can be seen on top of an 'underlayer' of densely
packed non-adipocytic cells. Original magnification ×40.

Long-term maintenance and subculture of human bone-derived cells

Passaging cells cultured in the absence of ascorbate

Materials

Ca^{2+} and Mg^{2+} free phosphate buffered saline (PBS)

Trypsin-EDTA solution (0.05% trypsin and 0.02% ethylenediaminete-traacetic acid in Ca^{2+} and Mg^{2+} free PBS, pH 7.4; GIBCO/BRL, Cat. No. 35400-027)

1 μg/ml DNAse I (Sigma Cat. No. D-4153)

70 μm cell strainer (Becton Dickinson, Cat. No. 2350)

50 ml polypropylene tube

Trypan blue solution (0.4% trypan blue in 0.85% NaCl, Sigma Aldrich Chemical Co. Ltd, Cat. No. T 8154)

Neubauer haemocytometer (Improved Neubauer BS 748; Merck, Cat. No. 403/0024/00)

80 cm^2 flasks

Narrow bore 2 ml pipette

Remove and discard the spent medium. Gently wash the cell layers 3× with 10 ml of PBS. To each flask, add 5 ml of freshly thawed trypsin-EDTA solution at room temperature (20 °C). Gently rock the flask to ensure that the entire surface area of the flask and explants is exposed to the trypsin-EDTA solution. Incubate the cells for 3–5 minutes at 37 °C. Remove the flasks from the incubator and examine under the microscope. Look for the presence of rounded, highly refractile cell bodies floating in the trypsin-EDTA solution. If none, or only a few, are visible, tap the base of the flask in an effort to dislodge the cells. If this is without effect, incubate the cells for a further 2 minutes at 37 °C and re-examine.

When the cells have become detached from the culture substratum, they are transferred to a 50 ml polypropylene tube containing 5 ml of FCS to inhibit tryptic activity. Wash the flask two to three times with 10 ml of serum-free medium (SFM) and pool the washings with the original cell isolate. To recover the cells centrifuge at 250 *g* (maximum) for 10 minutes at 15 °C. Aspirate the supernatant, invert the tube and drain briefly. Holding the top of the tube, sharply flick the base of the tube with the index finger to dislodge and break up the pellet. Add 2 ml of SFM containing 1 μg/ml DNAse I for each 80 cm^2 flask treated with trypsin-EDTA and using a narrow bore 2 ml pipette repeatedly aspirate and expel the medium to generate a cell suspension. The DNAse I is prepared as a 100 μg/ml stock in serum-free DMEM which is then filter-sterilised and stored frozen in small aliquots at −20 °C. Use once and discard.

Filter the cell suspension through a 70 μm 'cell strainer' to remove any bone spicules or remaining cell aggregates. The filters have been designed to fit into the neck of a 50 ml polypropylene tube. Wash the filter with 2–3 ml of SFM containing DNAse I and add the filtrate to the cells.

Take 20 μl of the mixed cell suspension and dilute to 80 μl with SFM. Add 20 μl of trypan blue solution, mix and leave for 1 minute before counting viable (round and refractile) and non-viable (blue) cells in a Neubauer haemocytometer. The number of viable cells/ml of suspension equals the number of viable cells per large square (in the centre or at the four corners of the grid) × the dilution factor (in this case 5) ×10^4. Using this procedure, typically 1×10^6–1.5×10^6 cells are harvested per 80 cm^2 flask of which ≥ 75% are viable.

Plate the harvested cells at a cell density suitable for the intended analysis. In the authors' experience, the *minimum* plating density for successful subculture is 3500 cells/cm^2. Below this, the cells exhibit extended doubling times and often fail to grow to confluence. In practice, the authors routinely subculture at 5×10^3–5×10^4 cells /cm^2 and achieve plating efficiencies measured after 24 h of ≥ 70%.

Passaging cells cultured in the presence of ascorbate

Materials

Collagenase (Sigma type VII from *Clostridium histolyticum*; Cat. No. C-0773 or Sigma type IV, Cat. No. C5138)
Calcium (CaCl$_2$, Merck, Cat. No. 19046 4K)

Due to their synthesis and secretion of an extensive collagen-rich, extracellular matrix, HBDC cultured in the continuous presence of ascorbate cannot be subcultured using trypsin-EDTA alone. They can, however, be subcultured successfully if first treated with purified collagenase. The procedure is as follows.

Rinse the cell layers twice with SFM (10 ml/80 cm^2 flask) and then incubate for 2 h at 37 °C in 10 ml of SFM containing 25 U/ml purified collagenase (Sigma type VII) and 2 mM additional calcium (1:500 dilution of a filter-sterilised stock solution of 1 M CaCl$_2$). Gently agitate the flask for 10–15 seconds every 30 minutes. The collagenase is prepared as a stock solution of 2500 U/ml stock in serum-free DMEM containing 5mM $CaCl_2$. Filter-sterilise and store frozen in small aliquots at -20 °C. Use once and discard. Sigma Type VII collagenase is highly purified and exhibits no detectable activity against non-collagen protein (R. Gundle, unpublished observation). Because of this, it is expensive, although a discount can usually be obtained if large quantities are purchased. As an alternative, the authors recommend the use of Sigma Type IV, which although less pure gives similar results and does not appear to compromise cell viability.

Terminate the collagenase digestion by discarding the medium (check, there should be no evidence of cell detachment at this stage). Gently rinse the cell layer twice with 10 ml of Ca^{2+}- and Mg^{2+}-free PBS. To each flask, add 5 ml of freshly thawed trypsin-EDTA solution (p. 59) at room temperature (20 °C). From this point on cells are treated as described above for cultures grown in the absence of ascorbate (pp. 59–60).

Typically, this procedure yields $\sim 3.5 \times 10^6$–4×10^6 cells/80cm^2 flask after 28–35 days in primary culture. Cell viability is generally $\geq 90\%$.

Induction of matrix mineralisation

Materials

100 μM L-ascorbic acid 2–phosphate
200 nM hydrocortisone or 10 nM dexamethasone
500 mM solutions of $Na_2HPO_4 \cdot 12H_2O$ and $NaH_2PO_4 \cdot 2H_2O$
25 cm^2 flasks
Cell scraper (Becton Dickinson, Cat. No. 3086)
Stain for alkaline phosphatase activity (Sigma staining Kit 86-R)
Haematoxylin Gill's No. 3
Von Kossa reagent
Substances for embedding for histology (see below)

Explants of human trabecular bone are prepared as described above (pp. 53–56) and cultured in medium supplemented with 100 μM L-ascorbic acid 2-phosphate and either 200 nM hydrocortisone or 10 nM dexamethasone. For studies of *in vitro* mineralisation, it is preferable to obtain trabecular bone from sites containing haematopoietic marrow. In practice, this is usually from the upper femur or iliac crest.

When the cells have attained confluence and synthesised a dense, extracellular matrix, typically after 28–35 days, subculture the cells using the sequential collagenase/trypsin-EDTA protocol (pp. 60–61) and plate the cells in 25 cm^2 flasks at a density of 10^4 viable cells/cm^2. Change the medium twice weekly.

After 14 days, supplement the medium with 5 mM inorganic phosphate (Pi). This is achieved by adding 0.01% (v/v) of a 500 mM phosphate solution, pH 7.4 at 37 °C, prepared by mixing 500 mM solutions of $Na_2HPO_4{\cdot}12H_2O$ and $NaH_2PO_4{\cdot}2H_2O$ in a 4:1 (v/v) ratio. The filter-sterilised stock solution can be stored at 4 °C.

After 48–72 h, the cell layers are washed two to three times with 10 ml of serum-free medium before fixation with 95% ethanol at 4 °C. This can be done *in situ*, for viewing *en face*, or, if sections are to be cut, following detachment of the cell layer from the surface of the flask using a cell scraper. Great care is needed if the cell layer is to be harvested intact, particularly when mineralised. Following fixation overnight at 4 °C, the cell layers are embedded in 2-hydroxymethylmethacrylate by the method described below, (references in Gundle,1995; Gundle, Joyner & Triffitt, 1995).

The embedding procedure should be carried out in a fume cupboard.

Solution A

2-hydroxymethylmethacrylate	94 ml
2-butoxyethanol	6–8 ml
Benzoyl peroxide (dried)	0.27 g

Solution B

Polyethylene glycol 400	30 ml
N,N-dimethylaniline	2 ml

Protocol

1 Dehydrate to absolute ethanol, then leave for 1 hour
2 Impregnate in solution A for 3 days at 4 °C
3 Place in a plastic mould and add a sufficient volume of premixed solution A and B (42 : 1 v/v) to cover.
4 Place in copper gas-tight container, to aid heat dissipation, containing phosphorus pentoxide as a desiccant. Evacuate the container, seal and allow the methacrylate to polymerise for 24 h at 4 °C.
5 Remove from the mould and mount on an aluminium block using Araldite™. Section at 5 μm using a tungsten carbide-tipped knife.

For the demonstration of alkaline phosphatase activity a staining kit (Sigma 86–R) is used. Staining solution (2.5 ml) is used per flask or sufficient to cover the section. Specimens are then placed in an humidified chamber and incubated for 1 h at 20 °C in the dark. The specimens are then washed under running tap water, and the nuclei counterstained for 15 seconds with haematoxylin. Mineral deposits are then stained using Von Kossa's technique. For examination under the microscope, sections are mounted in DPX, and the cell layers in flasks covered with glycerol.

Results

HBDC cultured in the continuous presence of glucocorticoid and the long-acting ascorbate analogue produce a dense extracellular matrix that mineralises extensively following the addition of Pi. This is the case for the original cell population grown from the explant at first passage (E1P1), and that obtained following replating of the trabecular explants (E2P1), which further attests to the phenotypic stability of the cultured cells (p. 64).

Cells cultured in the continuous presence of ascorbate and treated with glucocorticoid at E1P1 show only a localised and patchy pattern of mineralisation, despite possessing similar amounts of extracellular matrix and alkaline phosphatase activity (Gundle, 1995). Cells cultured without ascorbate, irrespective of the presence or absence of glucocorticoid, secrete little extracellular matrix and do not mineralise.

The ability of the cells to mineralise their extracellular matrix is dependent on ascorbate having been present continuously in primary culture. The addition of ascorbate in secondary culture, even for extended periods, cannot compensate for its omission in primary culture. This finding provides further evidence to support the hypothesis that maintenance of adequate levels of ascorbate during the early stages of explant culture is of critical importance for the survival of cells that retain the ability to proliferate extensively and give rise to precursors capable of undergoing osteogenic differentiation (Gundle, 1995).

The mineralised tissue formed by HBDC in vitro has yet to be characterised to the same extent as that formed in cultures of animal bone-derived cells (Aubin & Herbertson, Chapter 6) and cannot therefore, be defined as bone. That its formation is indicative of the true osteogenic potential of the cultured cell population is supported, however, by the recent demonstration that HBDC cultured in the continuous presence of ascorbate and glucocorticoid are capable of forming a tissue histologically resembling bone when implanted in vivo within diffusion chambers in athymic mice (Gundle et al., 1995; Gundle & Beresford, 1995).

Phenotypic stability

As a matter of routine our studies are performed on E1P1cells. Other investigators have studied the effects of repeated subculture on the phenotypic stability of HBDC and found that they rapidly lose their osteoblast-like characteristics. In practical terms, this presents real difficulties as it is often desirable to obtain large numbers of HBDC from a single donor.

As an alternative to repeated subculture, the potential of replating the trabecular explants at the end of primary culture into a new flask (E2) has been investigated. Using this technique, it is possible to obtain additional cell populations (E2P1) that continue to express osteoblast-like characteristics, including the ability to mineralise their extracellular matrix (pp. 61–63), and maintain their cytokine expression profile (Gallagher, *et al.*, 1996). For bone obtained from the majority of donors, although it is possible to continue culture through to the E3P1 stage, as a general rule, the authors do not use cells later than E2P1.

Cryopreservation

HBDC can be stored frozen for extended periods in liquid nitrogen or in ultra-low temperature (-135 °C) cell freezer banks. For this purpose, the cells are removed from the flask using trypsin-EDTA or collagenase-trypsin as appropriate, and following centrifugation resuspended ($1 \times 10^6 - 2 \times 10^6$ cells/ml) in a solution of (v/v) 50% medium, 40% serum and 10% dimethyl-sulphoxide. For best results, the cells should be frozen gradually using one of the many devices available that allow the rate of cooling to be controlled precisely. The authors have found the following freezing protocol to give excellent results:

5 °C /min $\Rightarrow$ 4 °C
1 °C /min $\Rightarrow$ -30 °C
2 °C /min $\Rightarrow$ -60 °C

The cells are then transferred directly to liquid nitrogen. Prior to use, frozen cells are rapidly thawed in a 37 °C water bath and then diluted into ≥ 20 volumes of preheated medium containing 10% FCS and the usual supplements. After 12–18 h, the medium is replaced with the normal volume of fresh medium. The efficiency of plating obtained after 24 h using this method is typically $\geq 70\%$.

Acknowledgements

RG was supported by a Wellcome Research Leave Fellowship. KS and JS were funded by a grant from The Bayer Corporation Inc. The authors acknowledge the many and important contributions made to the development of the human bone-derived cell culture system by other investigators.

References

Bard, D.R., Dickens, M.J., Edwards, J. & Smith, A.U. (1974). Ultra-structure, *in vitro* cultivation and metabolism of cells isolated from arthritic human bone. *J.Bone Joint Surg.*, **56B**, 352–60.

Bennett, J.H., Joyner, C.J., Triffitt, J.T. & Owen, M.E. (1991). Adipocytic cells cultured from marrow have osteogenic potential. *J.Cell Sci.*, **99**, 131–9.

Beresford, J.N., Bennett, J.H., Devlin, C., Leboy, P.S. & Owen, M.E. (1992). Evidence for an inverse relationship between the differentiation of adipocytic and osteogenic cells in rat marrow stromal cell cultures. *J.Cell Sci.*, **102**, 341–51.

Beresford, J.N., Gallagher, J.A., Poser, J.W. & Russell, R.G.G. (1984). Production of osteocalcin by human bone cells *in vitro*. Effects of $1,25(OH)_2D_3$, $24,25(OH)_2D_3$, parathyroid hormone and glucocorticoids. *Metab.Bone Dis. Rel.Res.*, **5**, 229–34.

Beresford, J.N., Gallagher, J.A. & Russell, R.G.G. (1986). 1,25-dihydroxyvitamin D_3 and human bone derived cells *in vitro*: effects on alkaline phosphatase, type I collagen and proliferation. *Endocrinology*, **119**, 1776–85.

Beresford, J.N., Graves, S.E. & Smoothy, C.A. (1993). Formation of mineralized nodules by bone derived cells *in vitro*: a model of bone formation? *Am.J.Med.Genet.*, **45**, 163–78.

Bruder, S.P., Fink, D.J. & Caplan, A.I. (1994). Mesenchymal stem cells in bone development, bone repair, and skeletal regeneration therapy. *J.Cell Biochem.*, **56**, 283–94.

Gallagher, J.A., Beresford, J.N., McGuire, M.K.B. *et al.* (1983). Effects of glucocorticoids and anabolic steroids on cells derived from human skeletal and articular tissues *in vitro*. In *Glucocorticoid Effects and Their Biological Consequences*, ed. B. Imbimbo, & L.V. Avioli, pp. 279–92. New York: Plenum Press.

Gallagher, J.A., Gundle, R. & Beresford, J.N. (1996). Isolation and culture of bone-forming cells (osteoblasts) from human bone. In *Methods in Molecular Medicine: Human Cell Culture Protocols*, ed. G.E. Jones, pp. 233–62. Totowa, NJ: Humana Press Inc.

Gehron-Robey, P. & Termine, J.D. (1985). Human bone cells *in vitro*. *Calcif.Tiss.Int.*, **37**, 453–60.

Gundle, R. (1995). Microscopical and biochemical studies of mineralised matrix production by human bone-derived cells. DPhil. Thesis, Oxford University.

Gundle, R. & Beresford, J.N. (1995). The isolation and culture of cells from explants of human trabecular bone. Calcif. Tiss. Int., **56**, S8–S10.

Gundle, R., Joyner, C.J. & Triffitt, J.T. (1995). Human bone tissue formation in diffusion chamber culture in vivo by bone-derived cells and marrow stromal fibroblastic cells. Bone, **16**, 597–601.

Hata, R-I. & Senoo, H. (1989). L-Ascorbic acid 2-phosphate stimulates collagen accumulation, cell proliferation and formation of a three-dimensional tissue-like substance by skin fibroblasts. J. Cell Physiol., **138**, 8–16.

Marie, P.J., Lomri, A., Sabbagh, A.& Basle, M. (1989). Culture and behaviour of osteoblastic cells isolated from normal trabecular bone surfaces. In vitro Cell Dev. Biol., **25**, 373–80.

Mills, B.G., Singer, F.R., Weiner, L.P. & Holst, P.A. (1979). Long term culture of cells from bone affected with Paget's disease. Calcif. Tiss. Int., **29**, 79-87.

Osipian, I.A., Tchailachian, R.K., Garibian, E.S. & Aivazian, V.P. (1987). The treatment of non-union fractures, pseudarthroses and long bone defects by transplantation of autologous bone marrow fibroblasts grown in vitro and impregnated into spongy bone matrix. Orthop. Traumatol., **9**, 59–61.

Parfitt, A.M. (1990). Bone-forming cells in clinical conditions. In Bone, A Treatise, Vol. 1: The Osteoblast and Osteocyte, ed. B.K. Hall, pp. 351–429. Caldwell, New Jersey: The Telford Press.

Stewart, K., Screen, J., Jefferiss, C.M., Walsh, S. & Beresford, J.N. (1996). Co-expression of the Stro-1 antigen and alkaline-phosphatase in cultures of human bone and marrow-cells. J. Bone Min. Res., **11**, P 208.

Wergedal, J.E. & Baylink, D.J. (1984). Characterization of cells isolated and cultured from human trabecular bone. Proc. Soc. Exp. Biol. Med., **176**, 60–9.

Wong, M.M., Rao, L.G., Ly, H., Hamilton, L., Tong, J., Sturtridge, W., McBroom, R., Aubin, J.E. & Murray, T.M. (1990). Long-term effects of physiologic concentrations of dexamethasone on human bone-derived cells. J. Bone Min. Res., **5**, 803–13.

5

Marrow stromal adipocytes

Jeffrey M. Gimble

Introduction

Development

The presence of adipocytes in marrow can be traced back phylogenetically to those species where the bone marrow first appears as a haematopoietic organ. In mammals, the extent of bone marrow adipogenesis is age related. In the neonate, the marrow cavity is filled predominantly with 'red' haematopoietic marrow and adipocytes are rarely seen. Soon after birth, adipocytes can be found in the periphery of the axial skeleton; for example, in the distal digits and tail vertebrae. With advancing age, there is a gradual replacement of the haematopoietic marrow with 'yellow' fatty marrow, a process which accelerates after puberty. The extent of development of fatty marrow correlates with the overall size and surface area : volume ratio of the animal. Thus, in the relatively small adult mouse, fatty marrow is limited to the distal tibia and fibula and the tail vertebrae, whereas in man it is present in up to 50% of the skeleton. The bulk is found in the long bones of the appendicular skeleton and, by the third decade of life, ≥90% of the femoral marrow cavity is occupied by fat. The development with age of marrow adipose tissue results from an increase in both the number and size of adipocytes (see also Bianco & Riminucci, Chapter 2).

Changes in the volume of marrow adipose tissue have also been correlated with the development of temperature gradients within the body. As individuals grow in stature, the temperature in their extremities and distal skeleton drops relative to the core temperature of their thoracic and abdominal cavities, and it has been hypothesised that the development of a fatty marrow at these sites is a response to these changes (Gimble, 1990; Tavassoli, 1989). The recent demonstration that marrow adipocytes can express the mitochondrial uncoupling protein, which was originally identified in brown adipose tissue and which allows for the direct conversion of energy from ATP into heat, supports the possibility that they might participate in temperature regulation at the extremities (Marko et al., 1995).

Relationship to adipocytes at extramedullary sites

Outside of the bone marrow, most of the body's adipose tissue is categorised as 'white adipose tissue' (WAT), which serves primarily as a storage site for triglycerides. In response to a decrease in calorific intake, or an increase in metabolic demand, these are hydrolysed into free fatty acids and released into the bloodstream. As a result, the volume of adipocytes in the WAT decreases, whereas, under these same conditions, the volume of adipocytes in the bone marrow remains unchanged, suggesting that they are relatively unresponsive to lipolytic stimuli. Indeed, it is only in extreme circumstances, periods of starvation lasting several months, that there is evidence for the mobilisation of bone marrow lipid stores. The distinction between bone marrow and extramedullary adipocytes is maintained in culture. Thus bone marrow adipocytes, when compared with those from WAT, are relatively resistant to the lipolytic actions of insulin. These findings have been interpreted as indicating the existence of a fundamental difference between these related cell populations (Gimble, 1990; Tavassoli, 1989).

All newborn mammals possess 'brown adipose tissue' (BAT) around their vital organs and in their interscapular region (Cornelius, MacDougald & Lane, 1994; Smas & Sul, 1995). The BAT serves as a thermogenic organ, providing a non-shivering source of heat. The expression of the mitochondrial uncoupling protein by adipocytes present in BAT and in the bone marrow (Marko et al., 1995) suggests that these cell types may be related functionally.

Potential function(s)

The function of marrow adipocytes has yet to be established with certainty (Gimble, 1990;Tavassoli, 1989). A number of potential functions, which need not be mutually exclusive, have been proposed. First, the simplest hypothesis is that they perform a purely passive role as 'space fillers', their number and size varying in inverse proportion to the amount of red haematopoietic marrow that is present. Secondly, in pathological situations, for example, in bone fracture or severe blood loss, marrow adipocytes may provide a local and readily mobilised source of metabolic intermediates to subserve the needs of haematopoiesis and or skeletogenesis. Thirdly, marrow adipocytes may play a role in lipid metabolism. In some species, a high percentage of chylomicrons introduced into the bloodstream are cleared by marrow adipocytes rather than in the liver. Fourthly, it has been postulated that marrow adipocytes participate in the regulation of bone turnover and of haematopoiesis. This possibility has recently been the focus of much attention and will therefore be considered in greater detail.

The relationship between adipogenesis and osteogenesis

In vitro and *in vivo* studies suggest that a reciprocal relationship exists between adipogenesis and osteogenesis in the bone marrow. In all forms of osteoporosis, irrespective of the cause, the decrease in trabecular bone volume is associated with an increase in the volume of marrow adipose tissue. In culture of rat marrow stromal cells, and in the murine marrow stromal cell line BMS2, an increase in the differentiation of adipocytes occurs at the expense of a decrease in the differentiation of osteoblasts and vice versa (Dorheim *et al.*, 1993; Gimble *et al.*, 1995; Beresford *et al.*, 1992). Evidence that marrow adipocytes exhibit considerable plasticity of phenotype is suggested by the findings of several investigators (for review, see Tavassoli, 1989). However, the most direct evidence in support of this, to date, has come from the study of clonal populations of rabbit marrow stromal cells, where it was shown that adipocytic clones could be induced to lose their accumulated lipid, re-enter a proliferative phase and subsequently express osteogenic potential when implanted *in vivo* within diffusion chambers (Bennett *et al.*, 1991). Collectively, these data are consistent with the possibility that osteoblasts and adipocytes share a common precursor until a late stage in their development.

Marrow adipocytes may influence the activities of osteoblasts and their immediate precursors by the production of soluble mediators (Benayahu, Zipori & Wientroub, 1993). In man, marrow adipocytes express the cytochrome P450 enzyme aromatase, which converts circulating androgens into oestrogen (Frisch, Canick & Tulshinsky, 1980). Thus, they have the potential to increase the levels of this key osteotropic factor in the immediate bone marrow microenvironment.

The relationship between adipogenesis and haematopoiesis

In vivo studies suggest that a reciprocal relationship exists between haematopoiesis and adipogenesis. If animals are hypertransfused with red cells, erythropoiesis is suppressed and adipocytes begin to occupy an increased percentage of what is normally a 'red' marrow cavity. However, this phenomenon is transient. Large numbers of granulocytes appear in the newly converted 'yellow' marrow. This suggests that adipocytes may release granulopoietic cytokines. In contrast, animals can be made anaemic by repeated phlebotomy. In the face of this erythropoietic stimulus, an animal's 'yellow' marrow rapidly loses its adipocytes and reverts to a 'red' or haematopoietically active state (Gimble, 1990; Bianco & Riminucci, Chapter 2).

Many of the murine marrow stromal cell lines which support lymphopoiesis *in vitro* are anecdotally noted to be pre-adipocytic. Studies on the BMS2 line have shown, however, that this ability is independent of the state of adipocytic differentiation of the cells, at least with regard to B-cell lymphopoiesis (Gimble, 1990; Deryugina & Muller-Sieberg, 1993).

Osteoclasts, the cells responsible for the resorption of bone, share a common origin with cells of the monocyte/macrophage series. In common with other cells of the haematopoietic system, the differentiation of osteoclasts is dependent on interactions with cells of the marrow stromal system and their secretory products (extracellular matrix and/or cytokines). It is now recognised that these include marrow adipocytes and their immediate precursors. Preadipocytic cell lines cloned from neonatal mouse calvaria, MC3T3-G2PA6 and ST2, were shown to support osteoclastogenesis when co-cultured with mouse splenocytes to a greater extent than an osteoblastic cell line cloned from the same source (MC3T3-E1)(Udagawa et al., 1989). In part, these effects may be mediated by the third component of the complement pathway (C3), which is produced by adipocytes (Table 5.1) and has been shown to potentiate the actions of macrophage colony stimulating factor (M-CSF) on proliferation and osteoclastogenesis in culture of murine bone marrow (Sato et al., 1993).

Studies using the BMS2 cell line have revealed that, in the presence of 1,25-dihydroxyvitamin D3, preadipocytes and fully differentiated adipocytes support osteoclastogenesis in co-culture with murine bone marrow to a similar extent and at least as well as murine osteoblast-like cells. In contrast, mature adipocytes are less capable of supporting the growth and proliferation of non-adherent myeloid cells, possibly reflecting a decrease in the expression of secreted M-CSF (Table 5.1) (K.A. Kelly & J.M. Gimble, unpublished observations). Marrow adipocytes may also influence osteoclast-mediated bone resorption. *In vitro*, marrow adipocytes rapidly acidify their culture medium and it is known that low pH is an activator of osteoclast activity (Arnett & Spowage, 1994).

Table 5.1. *Pre-adipocyte and adipocyte expressed genes*[a]

Lipogenic and lipolytic
Phosphoenolpyruvate carboxykinase (PEPCK) ↑
Glycerol 3 phosphate dehydrogenase (G3PD) ↑
Fatty acid synthase (FAS) ↑
Pyruvate carboxylase ↑
Acetyl CoA carboxylase ↑
GLUT4 ↑
Hormone sensitive lipase (HSL) ↑
Lipoprotein lipase (LPL) ↑
aP2/422/Fatty acid binding protein ↑
Carbonic anhydrase III ↑

Transcriptional regulators of adipogenesis
CAAT/enhancer binding proteins α,β (C/EBP) ↑
Peroxisome proliferator-activated receptors α,β/δ,γ1,γ2 (PPAR) ↑
Hepatic nuclear factor 3 (HNF3)
ADD1/SREBP

Cytoskeletal and extracellular matrix
Decorin ↓
Laminin ↑
Collagen IV,VI ↑
Collagen I, III ↓
Tenascin ↓
Osteopontin ↓
Bone sialoprotein ↓
Osteocalcin ↓
Actin ↓
Tubulin ↓

Hormonal/cytokine receptor
Insulin receptor ↑
Growth hormone receptor ↑
β-Adrenergic receptors ↑
Oestrogen synthase/aromatase/CYP19 ↑
Bone morphogenic protein receptor/ALK-3
Interleukin 6 receptor/gp130
Tumour necrosis factor receptor
Interleukin 1 receptor
Transforming growth factor β receptor
Vitamin D3 receptor (VDR)
Peroxisome proliferator-activated receptor (PPAR) ↑
Glucocorticoid receptor (GR)
Oestrogen receptor (ER)
Retinoic acid receptor (RAR, RXR)

Table 5.1. (*cont.*)

Cytokines
Macrophage colony stimulating factor (M-CSF) ↓
Interleukin 6 ↔
Interleukin 7 ↔
Granulocyte-monocyte colony stimulating factor (GM-CSF)
Bone morphogenetic protein ↓
Transforming growth factor-β
Insulin-like growth factors
Secreted proteins
Complement C3 ↑
Adipsin/complement factor D ↑
Angiotensinogen ↑
Apolipoprotein E ↑
Transmembrane proteins
CD36 (Glycoprotein IV/Fatty acid transport protein) ↑
CD9

Note:
[a] Where known, changes in protein expression that accompany adipogenesis are indicated thus: ↑, induced; ↓, decreased; ↔, unchanged.

Potential research applications

Increased numbers of bone marrow adipocytes are observed in patients with osteoporosis and in those with aplastic anaemia. In both diseases, the pathophysiological significance of this altered pattern of stromal cell differentiation is unknown. The results of studies on the mechanism of bone marrow adipogenesis will clearly have potential impact on the treatment of these and other disease states. Therapeutic application of cytokines or synthetic agents may allow physicians to target the direction of bone marrow stromal cell differentiation. This approach may prove invaluable in preventing bone loss associated with ageing and in the treatment of anaemias associated with cancer chemotherapy.

Characterisation of marrow adipocytes

Morphology and ultrastructure

Among the terms used for the precursor cells of bone marrow adipocytes are the reticulum/adventitial cell, the Westen–Bainton cell and mesenchymal or stromal progenitor cell. In this chapter the adipocytic precursor will be referred to as a stromal progenitor cell, and it is postulated that these precursors progress through a number of stages of commitment as they differentiate (Cornelius et al.,1994). A fully mature adipocyte is classically described as a 'signet ring cell' on the basis of its morphology in situ (Tavassoli, 1989). These cells contain a single, lipid-filled vacuole, surrounded by a thin ring of cytoplasm and a peripherally located nucleus. They contain little rough endoplasmic reticulum or glycogen and are enclosed by a basement membrane containing few collagen fibrils. This distinguishes them from adipocytes at extramedullary sites, which possess an abundance of endoplasmic reticulum and glycogen, and an extracellular matrix rich in collagen fibres.

Electron microscopy has been used to examine the cellular structure of primary bone marrow adipocytes and differentiating 3T3-L1 cells (Novikoff et al., 1980). In the undifferentiated state, the 3T3-L1 cell displays a fibroblast-like morphology, with actin stress fibres throughout the cytoplasm. These stress fibres decrease in number during differentiation. The lipid sphere in the mature adipocyte is encased by endoplasmic reticulum. Mitochondria and catalase positive microperoxisomes are closely associated with the endoplasmic reticulum and lipid droplet. A similar constellation of organelles exists in the liver, except that peroxisomes replace the adipocyte's microperoxisome. However, unlike fat droplets in the liver, the adipocyte lipid is bounded not by a membrane, but by an ordered complex of microfilaments.

Mature bone marrow adipocytes are esterase and performic acid Schiff reagent positive and alkaline and acid phosphatase negative. The fat vacuoles are stained by Oil Red O and Sudan Black, which identifies their contents as neutral lipids. Lipid vacuoles in cultured stromal adipocytes can be visualised by staining with Oil Red O, using the following method.

Protocol for staining with Oil Red O

Periodate–lysine–paraformaldehyde (fixative solution)

1 Prepare stock solutions of 0.2 M L-lysine dihydrochloride and 0.1 M dibasic sodium phosphate. These are mixed at a 1 : 1 volume : volume ratio and adjusted to pH 7.4. This solution can be stored at 4 °C indefinitely.

2 Prepare an 8% paraformaldehyde solution by warming 10 ml distilled water to 80 ° to 90 °C, adding 100 µl 10 M NaOH and dissolving 0.8 g paraformaldehyde. If the paraformaldehyde does not go into solution

completely, add additional NaOH or re-warm. This solution must be freshly prepared on the day of use.

3 Mix 3 volumes of the lysine/phosphate solution (final concentration 75 mM lysine, 37.5mM sodium phosphate) to 1 volume 8% paraformaldehyde solution (final concentration 2%). Add 0.00214 g/ml of sodium periodate (final concentration 0.1 M). Adjust the final solution to pH 7.4.

Fixation of cells

1 Cells can be cultured in dishes or on slides. After removing the medium and washing with phosphate-buffered saline, immerse the cells in fixative solution for 15 minutes. Remove the fixative, wash with phosphate-buffered saline, and air dry.

Staining of cells

1 Prepare a saturated stock solution containing 0.5 g Oil Red O in 100 ml of 99% isopropanol. This is stable at room temperature indefinitely.

2 Prepare a working solution on the day of use by mixing 6 ml of the saturated stock solution with 4 ml of distilled water. After 5 minutes, filter the solution through No. 46 Whatman filter paper. Immediately prior to use, filter the solution through a 0.2 μm syringe filter.

3 Immerse the cells in the working solution for 15 to 20 minutes. Wash three times with distilled water. If desired, the cells can be counterstained with haematoxylin and eosin.

4 Under light microscopy, lipid vacuoles will now be stained bright red, relative to other cellular organelles.

Cell and molecular biology

In vivo studies of bone marrow fat have been restricted to examining a relatively static cell population. Initial *in vitro* studies dealt with this limitation by employing long-term bone marrow cultures to study the dynamics of adipogenesis. However, the heterogeneity of the adherent cell population in these cultures limited interpretation of the experimental results. For this reason, a number of laboratories have developed cloned stromal cell lines capable of undergoing adipocytic differentiation *in vitro*. Pre-adipocytic lines have been successfully derived from bone marrow, calvaria and extramedullary adipose tissue in mouse, rat and rabbit. However, only one human cell clone with pre-adipocytic characteristics has been described, MG63, a human osteosarcoma cell line (Nuttall, Olivera & Gowen, 1994).

Lines derived from the marrow stroma and from extramedullary adipose tissue have greatly facilitated the investigation of the molecular and biochemical changes which occur during the differentiation of adipocytes (Deryugina & Muller-Sieberg, 1993). Based on gel electrophoretic analysis of protein lysates, investigators estimate that adipocytes contain more than 100 proteins which can be induced, and many others which are decreased in abundance during adipogenesis (Smas & Sul, 1995; Cornelius *et al.*, 1994). In Table 5.1 a partial list of these is given, and many of them can now be categorised based on their function. Where it is known, increases or decreases associated with adipogenesis are indicated in Table 5.1.

Transcription factors

The transcriptional regulation of adipogenesis is an active area of research. Two families of DNA binding proteins are of major importance. These are encoded by the CAAT/enhancer binding protein (C/EBP) genes (α, β, δ) and peroxisome proliferator-activated receptor (PPAR) genes (α, γ, δ) (Smas & Sul, 1995; Cornelius *et al.*, 1994). The PPAR proteins are members of the steroid receptor gene super-family. Like other steroid receptors, specific ligands bind to the PPARs. These include chemical compounds known as thiazolidinediones, arachidonic acid metabolites, and long-chain fatty acids. When fibroblast-like cells are co-transfected with genes encoding members of both the C/EBP and PPAR families, they are converted into pre-adipocyte cells. In the presence of appropriate inducing factors, the cells will rapidly accumulate lipid droplets. However, the identity of the natural ligand for the PPARs remains to be determined.

Other transcription factors have been associated with adipogenesis (ADD1/SREBP, HNF3). These transcriptional proteins are expressed during the pre-adipocyte and immature adipocyte stages of development (Cornelius *et al.*, 1994). The SREBP factors present a unique model of a membrane-bound transcription factor. In resting cells, SREBP is inactive and is localised in the endoplasmic reticulum. Upon activation by cholesterol depletion, a protease cleaves the SREBP protein. The released amino-terminal peptide then translocates to the nucleus where it regulates transcription (Hua *et al.*, 1995).

Lipogenic and lipolytic proteins

Not surprisingly, a number of induced proteins are involved in lipid or glucose metabolism within the cytoplasm. Non-cytoplasmic proteins are also included in this category. The glucose transporter, GLUT4, is a transmembrane protein. The lipolytic enzyme lipoprotein lipase is a secreted protein and one of the earliest enzymatic markers of adipocyte commitment. The enzyme glycerol phosphate dehydrogenase (GPD) provides an essential substrate for triacylglycerol synthesis and may account for the esterase positive staining of mature adipocytes. The assay for GPD is relatively simple and a protocol is shown below.

Glycerol-3-phosphate dehydrogenase (GPD) assay

1 Adipocyte differentiation is accompanied by increased expression of the enzyme glycerol-3-phosphate dehydrogenase (GPD), which can be assayed spectrophotometrically. Although lipoprotein lipase enzyme activity can also serve as a measure of adipocyte differentiation, the GPD assay is simpler and is recommended for laboratories not wishing to invest in specialised equipment.

2 To assay, wash cells in phosphate-buffered saline and harvest in 4 volumes of the following solution at 4 °C : 50 mM Tris-HCl, pH 7.6, 1 mM EDTA, 1 mM 2-mercaptoethanol. Homogenise the cells with 10 strokes of a Dounce homogeniser and centrifuge for 5 minutes, 8000 g at 4 °C. Recover the supernatant and centrifuge for 60 minutes at 100000 g at 4 °C. The final supernatant is then used in the spectrophotometric assay. An aliquot of the same is removed for the determination of total protein.

3 An aliquot of the 100000 g supernatant corresponding to 40 to 50 μg of total protein is assayed in a volume of 500 μl in the presence of the following reagents: 100 mM triethanolamine-HCl, pH 7.5, 0.1 mM 2-mercaptoethanol, 2.5 mM EDTA, 0.12 mM NADH, and 0.2 mM dihydroxyacetone phosphate. The reaction is initiated by the addition of the dihydroxyacetone substrate. The absorbance at 340 nm is then monitored at 15-second intervals over a 3 to 5-minute period. Purified rabbit muscle GPD of known concentration is used to obtain a standard enzyme activity curve. The results are expressed as units of GPD activity per mg protein.

Cytoskeletal and extracellular matrix proteins

Stromal progenitor cells or preadipocytes express an abundance of extracellular matrix proteins, many of which are also expressed by cells of the osteoblast lineage, e.g. osteopontin, osteocalcin and collagen type I (Dorheim et al., 1993). All of these are down-regulated during adipogenesis, while the expression of basement membrane proteins, including collagen types IV and VI and laminin, is increased (Table 5.1). Decreased expression of cytoplasmic structural proteins, such as actin and tubulin, also occurs. This may correlate with the morphological conversion of the cell from a flattened, adherent fibroblast to a more rounded adipocyte.

Hormone and cytokine receptors

A wide variety of hormones, cytokines and growth factors have been shown to influence the differentiation of adipocytes. Their receptors can be classified according to the mechanism(s) of signal transduction that they employ. The cytoplasmic domain of the insulin receptor is a tyrosine kinase, whereas those of the TGF-β superfamily, including the bone morphogenic proteins (BMPs), function as serine/threonine kinases. The receptor complex for interleukin 6 and its related cytokines (IL-11, leukaemia inhibitory factor, oncostatin M) share a common transmembrane protein, gp130. By engaging gp130, these receptors activate a common cytoplasmic pathway involving the Janus activated kinase (JAK) and the signal transducers and activators of transcription (STAT) family of transcription factors. Steroids (vitamin D3, glucocorticoids, peroxisome proliferators, estrogen, retinoids) bind directly to a conserved family of proteins, which act as transcriptional regulatory proteins. These receptors are expressed by both pre-adipocyte and adipocyte cells. Several are induced with adipogenesis; these include the insulin, BMP, and peroxisome proliferator γ2 receptors.

Cytokines and growth factors

Marrow adipocytes contribute to the milieu of the bone marrow microenvironment through the production of various growth factors and cytokines (Table 5.1). For some of these, secreted M-CSF and BMP, adipocytic differentiation-related changes in expression have been reported. In the case of others, however, for example interleukins 6 and 7, the level of expression is unaffected by the transition from preadipocyte to adipocyte (Table 5.1).

Secreted and transmembrane proteins

Bone marrow adipogenesis is accompanied by increased expression of several complement and acute phase proteins. The role of these mediators in adipocyte function remains unknown, but implies that bone marrow adipocytes may contribute indirectly to the immune response.

Isolation and culture of marrow adipocytes

Source of cells

To date, it has not proved possible to isolate a pre-adipocytic cell line from human bone marrow (p. 74). For this reason, studies of human adipogenesis have been limited to the use of long-term culture of marrow cells obtained by aspiration, most frequently from the ribs and pelvis. The problem of cellular heterogeneity in the adherent layer of these cultures has recently been addressed by the use of lineage specific antibodies to deplete the initial isolate of haematopoietic cells and enrich for stromal precursors (Gronthos, Graves & Simmons, Chapter 3). Included within this population are cells capable of undergoing adipogenesis, but cell lines have not yet been developed. The difficulty of cloning a human pre-adipocytic cell line may reflect a fundamental difference in the culture requirements of human marrow stromal cells when compared with those of other species, or that the incidence of stromal progenitors is much lower in human bone marrow. Whatever the reason, for the foreseeable future, studies of human adipogenesis will require regular and continued access to marrow aspirates.

For the study of adipogenesis in primary culture, bone marrow from adult rather than juvenile or newborn experimental animals should be utilised. Animals from larger species are preferred because a greater percentage of their marrow cavity is occupied by adipocytes (p. 67). The same is true for the long bones of the extremities as compared to the ribs and thoracic vertebral bodies.

Manipulation of the animal's physiology immediately prior to sacrifice can also be used to increase the number of marrow adipocytes. As noted earlier (p. 69), in polycythemic mice, increased numbers of adipocytes transiently fill the inactive marrow cavity. Similarly, a variety of drug treatment protocols can be used to increase the number of marrow adipocytes. Thiazolidinediones are a new class of oral anti-diabetic agents that act as specific ligands for the adipogenic transcription factor, PPAR. Chloramphenicol and related antibiotics have been associated with the development of aplastic anaemia. When rats are treated with either of these classes of drug, their bones respond with an increase in their content of 'yellow' marrow. Alternatively, the marrow content of stromal progenitor cells can be increased prior to harvest by treatment of the animal with chemotherapeutic drugs, for example, 5-fluorouracil, which rapidly depletes the marrow of lympho-haematopoietic cells.

Methods of isolation

Various sites can be used to isolate bone marrow stromal progenitor cells. The femoral bone marrow offers several advantages. In laboratory animals, the bone is readily accessible and the marrow cavity can be flushed out under sterile conditions. The majority of the cells are immediately in suspension and free of clumps without any additional steps. This contrasts with procedures for isolating stromal progenitor cells from neonatal calvarium which require sequential digestions with collagenase. While the femoral bone marrow contains a great number of haematopoietic cells, these can be depleted by physical means. The stromal progenitor cells adhere rapidly to culture dishes (within 2 to 4 hours), while the majority of haematopoietic cells remain non-adherent over the first 24 hours in culture. Alternatively, the haematopoietic cells can be depleted using specific antibody-coated magnetic beads. A variety of methods have been reported for cloning of preadipocyte stromal cell lines. Two protocols that have proved suitable for isolating stromal progenitor cells from rodent bone marrow are given below.

Cloning of preadipocyte stromal cell lines

Protocol 1 (Adapted from C.E. Pietrangeli, P.W. Kincade, G. Smithson, K. Medina, Oklahoma Medical Research Foundation)

1 Isolate the tibia and femur under sterile conditions from 4 to 6-week-old mice or rats. Remove the ends of the bone with scissors and flush the marrow cavity using a 3 ml syringe and a No. 18–20 (rat) or No. 25–26 (mouse) gauge needle with Dulbecco's modified Eagle's medium (4500 mg/l glucose) supplemented with 20% fetal bovine serum, 100 U penicillin/ml, 100 μg streptomycin/ml, 1 mM sodium pyruvate and 50 μM 2-mercaptoethanol. Pellet the cells by centrifugation at 1000g for 5 minutes. Re-suspend and adjust the cell count to 5×10^5 cells/ml and plate in flasks. Incubate at 37 °C in a humidified atmosphere of 93% air, 7% CO_2. In certain strains of mice, the following additional supplements have improved cell cloning: 20% conditioned medium from an established stromal cell line, endothelial cell growth factor (30 μg/ml), fibroblast growth factor (4 ng/ml), epidermal growth factor (4 ng/ml), insulin (5 μg/ml) and/or transforming growth factor ß (0.1 ng/ml).

2 Cells are maintained in culture on a weekly feeding schedule until confluence (approximately 2 to 3 weeks). The cells are subcloned by limiting dilution in 96-well microtitre plates. (To deplete any non-adherent

Table 5.2. *Factors influencing adipogenesis in culture of marrow stromal cells*

Agonists	Antagonists
Glucocorticoids	Inflammatory cytokines (TNF, IL-1)
Methylisobutylxanthine (IBMX)	gp130 receptor cytokines
	(IL-6, IL-11, LIF, Oncostatin M)
Indomethacin	Transforming growth factor-β family
	(TGF-β, BMPs)
Thiazolidinediones (Pioglitazone, BRL 49653)	1,25-Dihydroxyvitamin D3 (Calcitriol)
Foetal bovine serum (≥20% v/v)	
Biotin	
Insulin[a]	
1,25-Dihydroxyvitamin D3 (Calcitriol)	
Growth hormone	
Prolactin	
Thyroxine	

Note:
[a] Weak compared with its activity on preadipocytes derived from WAT.

haematopoietic progenitors, some investigators have treated the cultures with 50 µg/ml 5-fluoracil for 3 days prior to subcloning, but the necessity of this step is debatable.) Supplementation with 20% stromal cell-conditioned medium (sterile filtered to ensure that no cell contamination occurs) will improve the cloning efficiency. Cultures derived from a single cell are grown to confluence and expanded to duplicate plates.
3 Cells are examined under phase contrast microscopy to distinguish the adipocyte phenotype. In some clones, adipogenesis may occur spontaneously in the culture medium. The process can be accelerated by treatment of the cells with adipogenic agonists (Table 5.2).

Protocol 2 (Adapted from M.R. Candelore, Merck Research Laboratories)

1 Marrow from 4–5-week-old rat femur is aspirated into growth medium, collected by centrifugation (1000g), washed three times in phosphate buffered saline, and cultured in Dulbecco's modified Eagle's medium (1000 mg/l glucose) supplemented with 10% fetal bovine serum and 1 mM sodium pyruvate at 37 °C in 5% CO_2. After 4 days, the medium is supplemented with 20% human umbilical vein endothelial cell conditioned medium for 1 month; this improves the cloning efficiency of preadipocyte stromal cells exhibiting an epithelioid-like morphology.

2 Empirical observations suggest that bone marrow fibroblasts are readily removed from the tissue culture dish by treatment with 0.05% trypsin/0.53 mM EDTA for 1–3 minutes. The remaining adherent population is enriched for adipocyte precursors. Further enrichment can be achieved by Ficoll gradient centrifugation. Adherent cells (3–5×10^6/ml) are layered on a gradient of 3 ml 1:1 lymphocyte separation medium (Organon Teknika Corp Durham NC: 6.2 g Ficoll and 9.4 g sodium diatrizoate per 100 ml: density 1.077–1.080 g/ml): DMEM over 3 ml of undiluted lymphocyte separation medium. After 30 minutes of centrifugation (2400 rpm, room temperature), the cells banding at the lower interface region of the gradient are replated in growth medium.

3 After several enrichment cycles, a pure population of adipocyte precursors can be obtained. Cells are maintained in DMEM (1000 mg/l glucose), 2 mM L-glutamine, 110 mg/litre sodium pyruvate, 100 U penicillin/ml, 100 μg streptomycin/ml, 2.5% heat inactivated bovine serum, human acidic fibroblast growth factor (2.5 ng/ml) and heparin (5 μg/ml).

Mature adipocytes are less adherent than their immediate precursors; their content of low density lipid making them relatively buoyant. To isolate large numbers of these cells for further investigation, an adipocyte-rich bone marrow aspirate can be centrifuged through a Ficoll density gradient (Protocol 2 above). The adipocytes will be found floating above the gradient phase. Alternatively, the bone marrow aspirate can be placed in a sealed 75 ml culture flask that has been completely filled with culture medium. The adipocytes will float and adhere to the upper surface of the flask. After several days in culture, the adipocytes will have produced sufficient extracellular matrix to allow the flask to be inverted for conventional culturing.

Supplementation of the culture medium with 10–20% conditioned medium from an established stromal cell line will facilitate the isolation and growth of the adipocyte progenitors. Before purchase, the fetal bovine serum should be assayed to test for its ability to support adipogenesis.

Establishment of cultures

Adipogenesis is dependent on the presence of adipogenic agonists (Table 5.2) and of gluconeogenic substrates in the culture medium. The medium should contain a high content of glucose (4.5 mg/ml); Dulbecco's modified Eagle's medium is routinely used. This is supplemented with fresh sodium pyruvate, a limiting substrate in the gluconeogenic/glycerogenic pathway.

In most laboratories, marrow cells are cultured in medium containing fetal bovine serum at concentrations of 10–30%. Individual batches of fetal bovine serum vary in their content of adipogenic agonists and antagonists (Table 5.2). Pro-inflammatory cytokines and members of the TGF-β family of cytokines are potent inhibitors of adipocyte differentiation, even in the presence of strong agonists (Gimble *et al.*, 1995). Prior to purchasing a batch of fetal bovine serum therefore, it is advisable to test a sample using an established pre-adipocyte model.

In primary bone marrow cultures and in some cell lines, adipogenesis will occur spontaneously. However, the rate and extent of differentiation can be accelerated by the addition, at confluence, of agonists to the cells for 48 to 72 hours. There are a number of protocols for the induction of adipogenesis in culture of marrow stromal cells. Two of these are presented below.

Induction of adipogenesis in culture of marrow stromal cells

Protocol 1

1 Culture primary bone marrow stromal cells or a stromal cell clone to confluence. The cells should be relatively quiescent (less than one mitotic figure per field at ×20 magnification). The medium should contain a high glucose concentration (4.5 g/l) and be supplemented with fresh sodium pyruvate.

2 Prepare a stock concentration of hydrocortisone at 5×10^{-4} M in medium. This can be stored for periods of over 1 year in frozen aliquots ($-70\ °C$).

3 On the day of induction, weigh out the following chemicals:

1-Methyl-3-isobutylxanthine(IBMX)	16 mg
Indomethacin	3 mg

While these are poorly soluble in medium, they dissolve easily in 150 μl of dimethyl sulphoxide (DMSO). (If necessary, these compounds can be brought into solution with the addition of sodium hydroxide, but this necessitates a subsequent adjustment of the pH.) The IBMX is stable for periods of up to 3 months as a solid when kept at $-20\ °C$. Indomethacin is stable for periods of 1 year at room temperature.

4 Prepare the induction medium by diluting both the hydrocortisone and IBMX/indomethacin stocks 1:1000 into the culture medium (final concentrations 5×10^{-7} M hydrocortisone, 500 μM IBMX, 60 μM indomethacin). Incubate the cells with the induction medium for 3 days. They are then returned to their standard culture medium. By day 4 after

induction, small lipid vacuoles should be visible under phase contrast microscopy. These vacuoles will reach maximum size 6–7 days post induction. Mature adipocytes rapidly acidify their culture medium and require a change of medium every other day.

Protocol 2

Thiazolidinedione compounds can serve as alternative adipogenic agents. Prepare stock solutions in DMSO of either pioglitazone (MW 356.4) at 25 mM or BRL49653 (5-(4-(N-methyl-N(2-pyridyl)amino)ethoxy) benzyl)thiazolidine-2-4-dione; MW 357.4) at 5 mM. Use either of these reagents at a 1:1000 dilution in medium as an adipogenic agonist (final concentrations 25 or 5 μM, respectively). Maintain the cells for 3 days in the induction medium as described above.

In murine culture systems, the maximum numbers of adipocytes are obtained within 6 days of induction. The numbers of adipocytes can be determined using flow cytometry and the lipophilic, fluorescent dye, Nile Red (Smyth, Sparks & Wharton, 1993), for example, by the following method.

Fluorescence-activated cell sorting of adipocytes

1 Prepare a 1 mg/ml stock concentration of Nile Red in DMSO. This can be stored for several months at room temperature.
2 Harvest cultures by incubating with 0.25% trypsin/1mM EDTA and washing with phosphate buffered saline. Resuspend the cells at a concentration of 1×10^6–2×10^6 cells/ml and fix by the addition of paraformaldehyde to a final concentration of 0.5%. It is extremely important to digest the cell matrix completely and release the individual cells. If cell clumps are present they will cause blockages during the analysis. They should therefore be removed prior to fixation by passage through a 120 μm filter.
3 Immediately after fixing the cells, prepare a 1:100 dilution of the stock Nile Red solution in phosphate-buffered saline (10 μg/ml). Add 100 μl of the diluted Nile Red solution per ml of fixed cells (final concentration 1 μg/ml).
4 FACS analysis can be performed using a FACScan (Becton-Dickinson, San José, CA) multiparameter flow cytometer. A population of pre-adipocytic cells is used routinely as a negative control. An emission wavelength of 488 nm is used for excitation of the Nile Red. The FL2 detector,

which employs a bandpass filter of 585/42, is used to measure the gold fluorescence emission indicative of lipid vacuoles; this essentially monitors the wavelengths between 564 nm and 604 nm. The number of cells with enhanced gold fluorescence are determined in samples of 10^4 cells. Routinely, the adipocyte population exhibits an increased granularity relative to that of pre-adipocytes; this correlates with increased orthogonal light scattering.

If a flow cytometer is not available, numbers of adipocytes in stromal cell populations stained with the fluorescent dye, Nile Red, can be quantified by direct examination using a haemocytometer and fluorescence microscope (Zeiss research microscope with epifluorescence using Zeiss Filter Set 15). The total cell population can be visualised by staining with the DNA binding dye Hoechst 33342, or by determining the total cell number in a field by phase contrast microscopy. With both of the above methods, the percentage of cells exhibiting an adipocyte phenotype can be accurately determined. Alternatively, the number of adipocytes can be determined by staining with Oil Red O and phase contrast microscopy (pp. 73–74), although this is less efficient and accurate than determination by flow cytometry.

Long-term maintenance and subculture

Bone marrow adipocytes can be maintained in culture for up to several weeks without passage. However, within 24 to 48 hours, media in mature adipocyte cultures will turn yellow, indicating acidification; possibly due to the release of free fatty acids. Maintenance of the cultures therefore, requires frequent changes of medium (every 2–3 days). Continued culture in the presence of high levels of glucose is essential. In its absence, shrinkage of the fat vacuoles occurs, visible by phase contrast microscopy.

Mature adipocytes are viewed as a committed cell type, which is programmed to undergo few, if any, divisions (Cornelius et al., 1994). It is therefore advisable to subculture clonal lines in their pre-adipocytic state. In the author's laboratory, cells are subcultured weekly by treatment for 3–5 minutes with 0.25% trypsin/0.1% EDTA at 37 °C, resuspended in medium at the appropriate cell concentration ($10^4 - 10^5$ cells/ml) and then transferred to 35 or 100 mm diameter dishes (2 and 12 ml, respectively) or to 75 ml sealed flasks (10 ml). Cells plated at the above concentrations do not require a change of medium during the first week. At the end of this period, cell lines with doubling times of 36 to 48 hours will have reached confluence and, if they are contact inhibited, will be quiescent. It is at this stage that the cells should be exposed to adipogenic agonists (Table 5.2) as described above. If cultures are maintained at confluence for periods of ≥ 2 weeks, the cells become much less responsive to adipogenic agonists and only a minority ($\leq 10\%$) will undergo differentiation.

Following repeated subculture, cloned pre-adipocytic cell lines lose their ability to respond to adipogenic agonists. It is advisable therefore, to maintain adequate frozen stocks of cells at a low passage number; typically ≤ 10. The mechanism responsible for this loss of function is unknown, and it is sometimes necessary to sub-clone the parent line in order to obtain cells with a high adipogenic potential. This is most readily achieved by limiting dilution in 96-well micro-titre plates, and testing for adipogenic potential in the presence of an appropriate cocktail of agonists (Table 5.2)..

Freshly isolated marrow stromal cells can be passaged up to a maximum of four or five times. It is recommended that these cells be utilised for assays of adipogenesis within 2 weeks of reaching confluence. The maintenance and sub-culture of these cells is similar to that described for the pre-adipocytic cell lines.

Cryopreservation

For cryopreservation, pre-adipocytic stromal cell lines are harvested with trypsin-EDTA (p. 84), washed once with fresh medium, and then re-suspended in 90% fetal bovine serum: 10% DMSO (v : v) at $>2\times 10^6$ viable cells/ml. One ml aliquots are placed in sterile cryogenic vials and stored in liquid nitrogen. Under these conditions, the cells remain viable for several years. It is strongly recommended that investigators store five or more vials on two separate occasions immediately after receiving a new stromal cell clone.

To re-establish the cell clone from a frozen vial, the aliquot is thawed at 37 °C for less than one minute and immediately diluted 1:20 in fresh medium. Improved cell growth can be achieved by raising the fetal bovine serum concentration to 20% (v : v). After 24 hours, the culture medium should be replaced to remove all traces of DMSO. The cells are then maintained as described above (p. 84).

Acknowledgements

The author thanks Drs L. Borghesi, P.W. Kincade, K. Medina, D. Osmond, and G. Smithson for their critical evaluation of this manuscript; Ms M. Candelore for the protocol for cloning cell lines; M. Anderson and the staff of the OMRF OASIS for editorial assistance; and K. Kelly, C.E. Robinson, X. Wu and the rest of the OMRF Immunobiology and Cancer Program for their continued support.

References

Arnett, T.R., Boyde, A., Jones, S.J. & Taylor, M.L. (1994). Effects of medium acidification by alteration of carbon dioxide or bicarbonate concentrations on the resorptive activity of rat osteoclasts. *J.Bone Min.Res.*, **9**, 375-9.

Benayahu, D., Zipori, D. & Wientroub, S. (1993). Marrow adipocytes regulate growth and differentiation of osteoblasts. *Biochem.Biophys.Res.Commun.*, **197**, 1245-52.

Bennett, J.H., Joyner, C.J., Triffitt, J.T. & Owen, M.E. (1991). Adipocytic cells cultured from marrow have osteogenic potential. *J.Cell Sci.*, **99**, 131-9.

Beresford, J.N., Bennett, J.H., Devlin, C., Leboy, P.S. & Owen, M.E. (1992). Evidence for an inverse relationship between the differentiation of adipocytic and osteogenic cells in rat marrow stromal cell cultures. *J.Cell Sci.*, **102**, 341-51.

Cornelius, P., MacDougald, O.A. & Lane, M.D. (1994). Regulation of adipocyte development. *Annu. Rev. Nutr.*, **14**, 99–129.

Deryugina E.I. & Muller-Sieburg, C.E. (1993). Stromal cells in long-term cultures: keys to the elucidation of hematopoietic development? *Crit. Rev. Immunol.*, **13**, 115–50.

Dorheim, M-A., Sullivan, M., Dandapani, V., Wu, X., Hudson, J., Segarini, P.R., Rosen, D.M., Aulthouse, A.L. & Gimble, J.M. (1993). Osteoblastic gene expression during adipogenesis in hematopoietic supporting murine bone marrow stromal cells. *J.Cell Physiol.*, **154**, 317-28.

Frisch, R.E., Canick, J.E. & Tulshinsky, D. (1980). Human fatty marrow aromatizes androgen to estrogen. *J. Clin. Endocrinol. Metab.*, **51**, 394–6.

Gimble, J.M. (1990). The function of adipocytes in the bone marrow stroma. *New Biologist*, **2**, 304–12.

Gimble, J.M., Morgan, C., Kelly, K., Wu, X., Dandapani, V., Wang, C-S. & Rosen, V. (1995). Bone morphogenetic proteins inhibit adipocyte differentiation by bone marrow stromal cells. *J.Cell Biochem.*, **58**, 393-402.

Hua, X., Sakai, J., Ho, Y.K., Goldstein, J.L. & Brown, M.S. (1995). Hairpin orientation of sterol regulatory element-binding protein-2 in cell membranes as determined by protease protection. *J. Biol. Chem.*, **270**, 29422–7.

Marko, O., Cascieri, M.A., Ayad, N., Strader, C.D. & Candelore, M.R. (1995). Isolation of a preadipocyte cell line from rat bone marrow and differentiation to adipocytes. *Endocrinology*, **136**, 4582–8.

Novikoff, A.B., Novikoff, P.M., Rosen, O.M. & Rubin, C.S. (1980). Organelle relationships in cultured 3T3–L1 preadipocytes. *J. Cell Biol.*, **87**, 180–96.

Nuttall, M.E., Olivera, D.L. & Gowen, M. (1994). Control of osteoblast/adipocyte differentiation in MG-63 cells. *J. Bone Min. Res.*, **9**, (1) A28.

Sato, T., Abe, E., Cheng, H.J., Mei, H.H., Katagiri, T., Kinoshita, T., Amizuka, N., Ozawa, H. & Suda, T. (1993). The biological roles of the 3rd component of complement in osteoclast formation. *Endocrinology*, **133**, 397-404.

Smas, C.M. & Sul, H.S. (1995). Control of adipocyte differentiation. *Biochem. J.*, **309**, 697–710.

Smyth, M.J., Sparks, R.L. & Wharton, W. (1993). Preadipocyte cell lines: models of cellular proliferation and differentiation. *J. Cell Sci.*, **106**, 1–9.

Tavassoli, M. (1989). Fatty involution of marrow and the role of adipose tissue in hemopoiesis. In *Handbook of the Hemopoietic Microenvironment*, ed. M. Tavassoli, pp. 157–87. Clifton NJ: Humana Press.

Udagawa, N., Takahashi, N., Akatsu, T., Sasaki, T., Yamaguchi, A., Kodama, H., Martin, T.J. & Suda, T. (1989). The bone marrow-derived stromal cell-lines MC3T3-G2/PA6 and ST2 support osteoclast-like cell-differentiation in cocultures with mouse spleen-cells. *Endocrinology*, **125**, 1805-13.

6

Osteoblast lineage in experimental animals

Jane E. Aubin and Alexa Herbertson

Introduction

Scope of the chapter

Understanding the origin and differentiation of osteoprogenitors (OPC) has been hampered by the fact that unequivocal criteria for their identification have not been established, that they are relatively rare cells and that insufficient information is available on the differentiation stages through which they progress in becoming mature osteoblasts. Given that direct identification has been difficult, indirect functional and colony assays have been used to determine OPC, based on their capacity to produce recognisable differentiated progeny (osteoblasts) capable of forming a tissue that histologically and biochemically resembles bone (Aubin, Turksen & Heersche, 1993).

Friedenstein and colleagues were the first to demonstrate that bone marrow stroma contains cells with the capacity to form bone when transplanted *in vivo* in a diffusion chamber (references in Owen, Chapter 1). More recently, conditions have been determined in which bone formation can occur and be manipulated *in vitro* in cell populations derived from bone and the marrow stroma. The aim of this chapter is to describe the methodology currently in use in the authors' laboratory for the investigation of the differentiation of OPC in primary cultures of animal bone and marrow cell populations.

Historical perspective

The first report that cells isolated from the bones of fetal or neonatal rodents were able to form a bone-like tissue *in vitro* was made in the early 1980s in a number of laboratories (for review, see Aubin, Turksen & Heersche, 1993).

An important advance was the recognition that for osteogenic differentiation to occur reproducibly, in addition to serum, it was necessary to supplement the culture medium with vitamin C (L-ascorbic acid), β-glycerophosphate (β-GP) and a physiological concentration of a glucocorticoid. Under these conditions of culture, it was observed that three-dimensional, nodular structures that mineralised, developed post-confluence. Histologically, it was shown that these resembled woven bone and that they contained viable cells with the morphological features of osteoblasts and osteocytes (Bellows *et al.*, 1986). It was shown subsequently that these same conditions of culture also supported the osteogenic differentiation of cells present in cultures of marrow stromal cells (Maniatopoulos, Sodek & Melcher, 1988).

Since its inception, the bone nodule assay has been applied to the study of osteogenic differentiation in cell populations derived from the bones and marrow of several species and from animals at different stages of development. In some of the experimental models, the number of nodules that form can be used as an approximate index of the number of OPC present in the original cell population that were capable of differentiating under the specific conditions of culture employed. This has been demonstrated for fetal rat calvarial cell populations by limiting dilution analysis. A linear relationship was observed between the number of cells plated and the number of nodules formed over a wide range of plating densities. Based on the assumption that the incidence of OPC follows a Poisson distribution (p. 105), it was calculated that they occur with a frequency of approximately 1 in 300 of the total cell population under the standard conditions of isolation and culture. The data obtained also supported the conclusion that nodule formation was a 'single-hit' phenomenon, i.e. that a single cell type was limiting for nodule formation in calvarial populations. Based on the data, it was concluded that a single initiating cell, referred to as a colony forming unit osteogenic (CFU-O), gave rise to all of the cells necessary for nodule formation (Bellows & Aubin, 1989).

In contrast to the findings made using the fetal rat calvarial model, in culture of stromal cells isolated from the bone marrow of young adult rats, the number of nodules formed, and hence CFU-O, does not show a linear relationship with the number of cells explanted, except at high plating density. A linear relationship is observed over all plating densities, however, if the limiting dilution analysis is performed in the presence of a fixed number of cells from the non-adherent (haematopoietic) fraction, endothelial cells or their conditioned medium. It was concluded that, in this system, except at high plating densities, nodule formation requires the co-operation of multiple, as yet ill defined, cell types and/or their secretory products (Aubin, Fung & Georgis, 1990). Observations made using cultures of murine marrow stromal cells are consistent with this possibility (Friedenstein et al., 1992).

OPC derived from fetal rat calvaria and from young adult rat bone marrow have a limited capacity for self-renewal when cultured in vitro (for review, see Aubin et al., 1993). Supplementation of the culture medium with low concentrations of a glucocorticoid (typically 10 nM dexamethasone) extends their functional life span and reveals the presence of a distinct population of nodule-forming cells (Bellows, Heersche & Aubin, 1990), which can be distinguished, and separated, on the basis of the cell surface expression of alkaline phosphatase (AP). The alkaline phosphatase-positive(AP+) OPC predominate in primary cultures of the calvarial-derived cell populations and are capable of forming nodules in the absence of added glucocorticoid. In contrast, alkaline phosphatase-negative(AP−) OPC predominate in the marrow-derived cell cultures and their nodule-forming capability is markedly enhanced by glucocorticoid even in primary cultures. The most likely explanation for this difference in glucocorticoid dependency is that the AP−OPC occur earlier in the hierarchy of osteoblast differentiation than the AP+OPC (Turksen & Aubin, 1991).

Although the use of indirect functional assays has made a significant contribution to our understanding of the nature of OPCs and the factors that influence their activities, they do have important limitations. Experimental approaches which rely on the identification of an 'end-stage' phenotype under a given set of culture conditions, do not normally allow discrimination of alternative pathways within the hierarchy, because not all stem or progenitor cells may realise their full potential for proliferation and differentiation. Moreover, since the end-stage phenotype may be several generations removed from the original, and in some cases multipotential precursor, it may not be possible to identify with certainty the factors that influence their commitment and determine their ultimate fate (Aubin et al., 1993).

Characterisation of cells of the osteoblast lineage

Morphology and ultrastructure

By the use of morphological, histochemical and biochemical criteria, four stages of the osteoblast lineage are evident in vivo: preosteoblasts, osteoblasts, lining cells and osteocytes. Preosteoblasts express AP activity and are located behind the osteoblast layer. Osteoblasts are commonly described as plump, basophilic cells containing a well-developed golgi apparatus, abundant endoplasmic reticulum and free ribosomes that are found at sites of bone formation, and that are actively engaged in the synthesis and secretion of a type I collagen-rich extracellular matrix. They are non-proliferating, strongly AP+ and express several non-collagenous proteins, including decorin, biglycan, osteonectin, osteopontin, bone sialoprotein and osteocalcin. The synthesis of some of these (for example, bone sialoprotein and osteocalcin) is highly restricted and within bone is essentially limited to mature cells of the osteoblast lineage (Aubin *et al.*, 1993).

Osteocytes are considered to be the terminally differentiated cell of the osteoblast lineage, and are derived from osteoblasts that become embedded within the bone matrix. They develop long, cellular processes that pass through canaliculi to make contact, via gap junctions, with other osteocytes as well as with the cells present on bone surfaces. They are AP− and express osteocalcin but not bone sialoprotein. It has been postulated that they participate in the regulation of mineral homeostasis and in the perception and transduction of mechanical stimuli (Van der Plas & Nijweide, 1992). Lining cells are flat, highly elongated and contain few of the intracellular organelles that characterise highly secretory cells. In the adult they are postulated to be present on essentially all quiescent bone surfaces. Although marker expression as outlined above is generally observed in most studies, there is a growing number of examples of species and site-specific marker expression patterns (for review, see Aubin & Liu, 1996).

Changes in gene expression associated with the differentiation of cells of the osteoblast lineage

The process of osteoblast differentiation *in vitro* recapitulates the *in vivo* stages. Analysis of the time course of cell proliferation within calvaria cell cultures in relation to the onset and completion of nodule formation and mineralisation has shown that genes associated with proliferative stages, e.g. histones, proto-oncogenes such as c-fos and c-myc, characterise the first phase, while certain cyclins B and E, are up-regulated post-proliferatively (Lian & Stein,1995). Following the cessation of proliferation, there is a progressive and sequential up-regulation of genes associated with the differentiated osteoblast phenotype: type I collagen, alkaline phosphatase, osteopontin, bone sialoprotein, (matrix maturation) and osteocalcin (matrix mineralisation). These changes in gene expression take place in parallel with osteogenic differentiation and the formation of nodules (Aubin *et al.*, 1993; Aubin & Liu, 1996). As noted above (pp. 88–89), the nodules resemble woven bone histologically and contain cells with the morphological and histochemical characteristics of osteoblasts, osteocytes and lining cells. A similar developmental progression has been reported to occur in cultures of adult rat marrow stromal cells (Malaval *et al.*, 1994).

The identification of cells of the osteoblast lineage earlier than the preosteoblast is a major challenge. Evidence suggests that they are present in the periosteum and in the bone marrow, and have fibroblastic morphology but do not possess any distinguishing morphological characteristics. The authors have recently begun to use the technique of replica plating to 'trap' OPCs prior to their overt morphological differentiation, in order to determine which clones will ultimately be capable of osteogenic differentiation (pp. 107–108) (Liu *et al.*, 1994). In addition to assaying for the expression of the osteoblast-related genes described above, the authors have used replica plating in combination with poly(A)-PCR to study the expression of the genes for potential regulatory growth factors and/or their receptors. These have included parathyroid hormone-related peptide (PTHrp) and its receptor, FGFR-1 and the α-subunit of the PDGF receptor (PDGF-α). The expression of each of these was found to be modulated during the process of osteoblast differentiation and a primitive population of cells identified that did not express the genes for alkaline phosphatase or type I collagen but did express those for PTHrp, FGFR-1 and PDGF-α (Aubin *et al.*, 1995; F. Liu & J.E. Aubin, unpublished results).

A corollary technique of 'cDNA fingerprinting' has also been developed to identify novel genes that are expressed only at specific stages of osteoblast differentiation. One of these, designated 1.1, codes for an mRNA of 750 bp that is highly expressed by newly differentiated osteoblasts at the onset of nodule formation. Sequence analysis, combined with a detailed search of several databases, revealed that this is a product of a novel gene that does not share significant homology with any previously cloned cDNA (Candeliere, Liu & Aubin, 1996). Based on findings to date, it is anticipated that these molecular approaches will add rapidly to the existing repertoire of stage specific markers for differentiating cells of the osteoblast lineage and thereby enhance our ability to study these cells *in vitro* and *in vivo*.

Antibodies recognising cells of the osteoblast lineage

The production of antibodies that recognise differentiation stage-specific antigenic determinants provides an additional means for the identification and isolation of cells of the osteoblast lineage (for recent review, see Aubin & Turksen, 1996). To date, the majority of antibodies generated have shown specificity for the more mature cells in the lineage (pre-osteoblasts,

osteoblasts and osteocytes) rather than their undifferentiated precursors. Of these, a number have been shown to recognise epitopes present on the bone/liver/kidney isoform of alkaline phosphatase, and these have proved useful for the immunoselection of osteoblasts and their immediate precursors. Antibodies showing specificity for osteoid osteocytes and/or osteocytes have now been generated in several laboratories (Walsh *et al.*, 1994; Van der Plas & Nijweide, 1992). A number of the antibodies produced recognise both early and late stage cells of the osteoblast lineage. Included in this category is the monoclonal antibody RCC455.4. This was generated in the authors' laboratory and found by expression cloning to recognise galectin 3 which is markedly up-regulated as precursor cells become mature osteoblasts and osteocytes (Aubin & Turksen, 1996).

Antibodies recognising cells of the osteoblast lineage earlier than the preosteoblast have proved difficult to obtain. Haynesworth and colleagues have reported the production of a series of monoclonal antibodies that react with a subset of human marrow stromal cells but which do not recognise osteoblasts or osteocytes. On this basis, these authors proposed that their antibodies reacted with antigenic determinants present on marrow stromal progenitors or even stem cells and that their expression is developmentally regulated (Haynesworth, Baber & Caplan, 1992).

Simmons *et al.* have described the production of a monoclonal antibody, STRO-1, that recognises a subset of human marrow stromal cells that includes all colony forming units-fibroblastic (CFU-F). Freshly isolated human marrow stromal cells, selected on the basis of the expression of STRO-1 and the lack of expression of glycophorin A, which eliminates STRO-1 positive erythroblasts, proliferate extensively and give rise to cells of multiple marrow stromal cell lineages, including osteoblasts, when cultured *in vitro*. On the basis of this and other evidence, it has been postulated that STRO-1 reacts with primitive, multi-potential marrow stromal precursors (Gronthos, Graves & Simmons, Chapter 3).

Isolation and culture of cells of the osteoblast lineage

Source of cells

Cells capable of forming mineralised nodules have been isolated from the bones and marrow of many species including man (for review see Beresford, Graves & Smoothy, 1993). The choice of animal model is driven by many considerations, some of which are historical, and each has its advantages and disadvantages. When attempting to isolate cells from bone, the age of the animal is important and will largely determine the method of isolation. The use of proteolytic enzymes (principally collagenase; pp. 94–95) works well with the calvarial bones of foetal or neonatal rodents and embryonic chicks. For the heavily mineralised bones of the adult, however, the use of explants, with or without a collagenase pretreatment, is the method of choice. For the isolation of marrow stromal cells (pp. 96–97) investigators have used the long bones of embryonic chicks, neonatal pigs, young adult mice, rats and rabbits and adult humans.

An additional consideration is the size of the animal: smaller animals are easier to house but greater numbers are required in order to recover sufficient cells for experimentation. For example, to seed one 75 cm^2 flask requires the bone marrow of three to four chick embryos but only one young adult rat. When using the porcine model, sufficient cells are obtained from each animal to seed four such flasks. The use of murine and human cell culture systems for the study of haematopoiesis has resulted in the widespread availability of purified and/or recombinant cytokines and numerous well-characterised antibodies recognising cell surface determinants (CD antigens) as well as cytokine receptors and their ligands. Thus, dependent upon the nature of the investigation, there may be a clear advantage in using cells of murine or human origin.

When selecting the source of cells, the potential existence of species differences should always be considered. An excellent example of a species difference is the glucocorticoid dependence of osteogenic differentiation in mice and rats. In rats, the effect of glucocorticoids is stimulatory in both calvarial and stromal cultures, whereas in mice it is inhibitory at least in stromal cultures (Aubin et al., 1993; Falla et al., 1993). Additional factors for consideration include the possible existence of age and/or sex-related differences in the behaviour of cells as well as differences related to their skeletal site of origin. This is an area of research that has been largely neglected, but there is a growing appreciation that it may have considerable bearing upon the design and interpretation of experiments (for discussion, see Aubin & Liu, 1996).

Methods of isolation

Foetal rat calvaria cells (FRC)

Materials

Collagenase 3 mg/ml (Sigma, St Louis, MO, USA Cat. No. C-0130)
DNAse 9.7 units/ml (Sigma Cat. No. D-4513)

Chondroitin sulphate 0.12 mM (Fluka Cat. No. 27043)
Sorbitol 100 mM
KCl 111.2 mM
$MgCl_2$ 1.13 mM
$ZnCl_2$ 0.5 mM
Glucose 13 mM
Tris-HCl 21.3 mM, pH 7.4
α-MEM supplemented with 15% fetal calf serum (FCS), 100 μg/ml
    penicillinG, 50 μg/ml gentamycin and 300 ng/ml fungizone
Trypsin 0.01% w/v in citrate saline pH 8.5 (Gibco Life Technologies, Inc.
    Gaithersburg, MD, USA Cat. No. 15090-038)
L-ascorbic acid 50 μg/ml
35 mm and 100 mm diameter Petri dishes
Electronic particle counter (Coulter Electronics, ModelZf, Hialea, FL, USA)

The following procedure is based on that originally described by Bellows *et al.* (1986). The calvarial bones from 21-day-old fetal rats are minced with fine scissors and then digested sequentially five times for 10 to 20 minutes at 37 °C in a solution comprising 3 mg/ml collagenase, 9.7 units/ml DNAse, 0.12 mM chondroitin sulphate, 100 mM sorbitol, 111.2 mM KCl, 1.3 mM $MgCl_2$, 0.5 mM $ZnCl_2$, 13 mM glucose and 21.3 mM Tris-HCl, pH 7.4. The released cells from each digest (populations I to V) are centrifuged at 1100 rpm, the supernatant removed, the cells resuspended in medium, plated separately and cultured at 37 °C in α-MEM supplemented with 15% fetal calf serum (FCS), 100 μg/ml penicillin G, 50 μg/ml gentamycin and 300 ng/ml fungizone in an humidified atmosphere of 95% air/5% CO_2. After 24 h, the cells are recovered by trypsinisation (0.01% w/v in citrate saline), and then counted using an electronic particle counter. Population I, a heterogeneous mixture containing predominantly undifferentiated fibroblastic cells, blood cells and macrophages, is usually discarded and the remaining cell populations (II–V) are pooled prior to further investigation.

After 24 h, and when the bones are digested in batches of about 25, the above procedure typically yields between 0.5×10^6–1×10^6 viable cells/calvaria. The average yield is reduced when the number of calvariae used per digestion is decreased. The presence of debris and non-viable cells precludes an accurate electronic determination of cell number in the original isolate. When the cells released in the successive digestions are not pooled immediately, the typical yields per population from 25 calvariae after 24 h are approximately 1×10^7–2×10^7, 5×10^6 and 5×10^5 viable cells for populations I, each of II – IV and V, respectively.

To obtain discrete colonies, the cells are subcultured at approximately 5–40 cells/cm² in 100 mm diameter Petri dishes. For most nodule forming assays, cells are subcultured into 35 mm Petri dishes at a density of $3.3 \times 10^3/\text{cm}^2$. To obtain a sufficient number of cells for the preparation of RNA, 100 mm Petri dishes and a plating density of $2 \times 10^3 – 8 \times 10^3/\text{cm}^2$ are required.

The cells are cultured as described above for periods of up to 35 days in medium additionally supplemented with 50 µg/ml L-ascorbic acid. The medium is changed after 24 h, at which time test substances may be added (pp. 97–99), and thereafter at 48 h intervals.

Rat marrow stromal cells (RMSC)

Materials

α-MEM supplemented with 10% FCS, 50 µg/ml L-ascorbic acid and anti-
biotics and antimycotics as on p. 95 (complete medium)
T-75 culture flasks
Dexamethasone 10 nM
Phosphate buffered saline, pH 7.4, with Ca^{2+} and Mg^{2+} (PBS)
Trypsin 0.25% w/v in citrate saline pH 8.5
Syringe with 22-gauge needle
100 mm Petri culture dishes

40 to 43-day-old (110–120 g) male Wistar rats are sacrificed by cervical dis-location. The femora are then removed under aseptic conditions and placed in α-MEM containing antibiotics and antimycotics as described in (p. 95). Following removal of the musculature and the adherent connective tissue, the femora are placed in fresh medium and the epiphyses and metaphyses removed with a scalpel. To obtain a suspension of marrow cells, the exposed femoral cavity is then flushed 5–8 times with 5 ml of complete medium. Typically, this procedure yields approximately 2×10^7 viable, nucleated cells from two femora.

The cells isolated from the bones of different rats are pooled and then cul-tured (2×10^7 cells per T-75 flask) in complete medium, plus or minus 10 nM dexamethasone (p. 98), at 37 °C in a humidified atmosphere of 95% air/5% CO_2. The medium is changed every second or third day. After 7 days, the cultures are washed with 15 ml of PBS pre-warmed to 37 °C to remove debris and any remaining non-adherent cells and then incubated for 1 minute in 5 ml of trypsin (0.25% w/v in citrate saline) at 37 °C. To the trypsin solution is then added 5 ml of collagenase solution (p. 95), and the incubation continued until the majority of cells are released (<30 min). Recovered cells are washed in medium, collected by centrifugation as above and resuspended in fresh medium. To disperse aggregates and ensure a uniform suspension, the recovered cells are again passed gently three to four times through a syringe with a 22 gauge needle.

The suspended cells are counted (typical recoveries are between 5×10^6 and 7×10^6 cells per rat) and then subcultured at densities between 10^3 and 5×10^4 cells/cm^2 into either 35 mm (nodule formation and/or colony-forming efficiency) or 100 mm (preparation of RNA or analysis of recovered cells by flow cytometry) diameter Petri dishes. They are then cultured in complete medium in the absence or presence of 10 nM dexamethasone and any other test substances for between 7 and 35 days. The exact period of culture will reflect the experimental endpoint (typically 18–21 days for nodule formation) and the number of cells required for further biochemical analysis.

When RMSC are isolated and cultured in medium containing dexamethasone, the incidence of fibroblastic colonies is approximately 1 in 50 adherent cells at first subculture, whereas for CFU-O it is approximately 1 in 300 (Aubin, Fung & Georgis, 1990). However, there is considerable variation in nodule forming efficiency between different RMSC isolates. For this reason, it is advisable always to use a range of plating densities within each experiment.

Factors influencing the formation and mineralisation of nodules

Serum supplement

There is a wide variation in the ability of different batches of FCS to support the formation of mineralised nodules in cultures of FRC and of RMSC. For this reason, it is essential that batches of serum are screened for their ability to support proliferation and nodule formation in both cell culture systems prior to purchase.

Vitamin C (L-ascorbic acid)

The formation of nodules is absolutely dependent on the maintenance of adequate levels of vitamin C, irrespective of the species of origin or the site of origin of the cultured cells (for review, see Beresford, Graves & Smoothy, 1993). Typically, this requirement is met by supplementation of the culture medium with 50–100 μg/ml L-ascorbic acid.

Glucocorticoids

Glucocorticoids exert a powerful influence on nodule formation and mineralisation. Within species, the degree of glucocorticoid dependency can depend on the skeletal site of origin and/or the developmental age of the cells under investigation. When added to cultures of FRC, the effect of glucocorticoids is stimulatory and reveals the presence of an additional class of nodule-forming cells (Bellows & Aubin, 1989, Bellows *et al.*, 1990, Turksen & Aubin 1991), whereas in cultures of RMSC the formation of nodules is essentially glucocorticoid dependent (Maniatopoulos *et al.*, 1988). When using the synthetic glucocorticoid dexamethasone to promote nodule formation, in cultures of both FRC and RMSC, the optimal dose is 10 nM and it should be added throughout primary and secondary cultures to maximise nodule numbers. In cultures of RMSC, however, this treatment protocol results in an approximately five-fold reduction in the number of cells harvested at the end of primary culture (Maniatopoulos *et al.*, 1988). Thus, the inclusion of dexamethasone in primary cultures should be considered optional and dependent upon the experimental objectives.

Inorganic phosphate concentration

For nodule mineralisation to occur, the culture medium must contain a source of inorganic phosphate (Pi). This can be achieved by the direct addition of 2–5 mM Pi (in the form of $NaH_2PO_4 \cdot 2H_2O$) or 10 mM sodium-β-glycerophosphate (β-GP) (Bellows, Heersche & Aubin, 1992). In cultures of FRC, the formation of mineralised nodules occurs typically after 18–21 days when β-GP is present continuously. In cultures of RMSC, however, these same conditions can result in the non-specific deposition of mineral in the cell layer, especially in areas where senescent cells reside. This can be avoided by delaying the addition of the β-GP until 2 days prior to fixation (p. 99).

Hormones, cytokines and growth factors

A wide variety of hormones, cytokines and growth factors have been shown to influence nodule formation in cultures of FRC and of RMSC and in the majority of cases the observed effects were inhibitory. The effects of TGF-β1, for example, are consistently inhibitory, irrespective of the timing and duration of exposure. In the case of others (epidermal growth factor, leukaemia inhibitory factor, 1,25-dihydroxyvitamin D3), biphasic effects are observed, dependent either on the timing of exposure, and hence the cells state of maturation, and/or the presence or absence of dexamethasone (triiodothyronine) (for discussion, see Aubin *et al.*, 1993).

Fixation and staining of nodules

Materials

2.5% silver nitrate in distilled water (von Kossa's reagent)
10% neutral buffered formalin (NBF); (100 ml formalin, 16 g. Na_2HPO_4,
    4 g. $NaH_2PO_4 \cdot H_2O$, distilled water to1 litre)
10 mg naphthol AS $MX-PO_4$ (Sigma)
400 µl *N,N*-dimethylformamide
0.1 M Tris-HCl pH 8.3
60 mg Red Violet LB salt (Sigma)
Whatman's No 1 filter
Toluidine blue (0.1% w/v in 30% v/v ethanol)

Bone nodules are identified by their characteristic morphology and, when mineralised, positive reaction with von Kossa's reagent. The histochemical demonstration of AP activity is frequently used in combination with von Kossa's reagent to demonstrate the presence of cuboidal AP+osteoblasts apposed to newly synthesised bone matrix (osteoid) as well as the presence of AP+cells at the nodule periphery (Fig. 6.1). It is important to appreciate, however, that the relationship between AP expression and mineralisation is not absolute (Fig. 6.2*a*), particularly in culture of RMSC (Herbertson & Aubin, 1995). For this reason, bone nodules should never be scored purely on the basis of AP expression.

    Cultures are rinsed once with PBS at 4 °C and then fixed in NBF for 15 min at 4 °C The fixative is removed and the cultures rinsed with distilled water (1×1 min and 1×15 min). The AP substrate (10 mg naphthol AS $MX-PO_4$ dissolved in 400 µl of *N,N*-dimethylformamide) is added to 99.6 ml of 0.1 M Tris-HCl, pH 8.3 and then mixed prior to the addition of 60 mg Red Violet LB salt. This solution is prepared freshly and then filtered immediately prior to its addition to the cultures. The reaction is allowed to proceed for 45 min at 20 °C. At the end of this period, the cultures are rinsed in tap water, drained and then stained for 30 min at room temperature with von Kossa's reagent. The cultures are then drained, rinsed twice in distilled water, counterstained with toluidine blue for 2 seconds, rinsed three times with tap water and allowed to air dry.

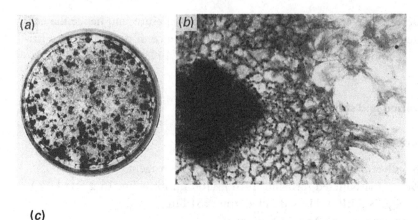

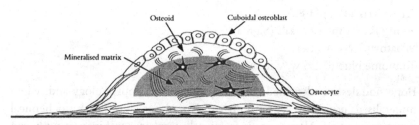

Fig. 6.1.(a). A 21-day culture of RMSC grown in the presence of ascorbic acid, β-GP, and dexamethasone. The culture has been stained for the presence of mineral (von Kossa's reagent) and of AP. Bone nodules appear black and areas of AP expression grey. Magnification ×1.6.

(b). Photomicrograph of a bone nodule formed in a 21-day culture of RMSC. Note the characteristic features of the nodule with a well-mineralised central region (black) and the presence of AP positive cells (grey) possessing a polygonal morphology at the periphery. Magnification ×100.

(c). Diagrammatic representation of a bone nodule in cross-section. Based on a Figure prepared by Dr L. Malaval (INSERM U403, Lyon, France).

For the determination of nodule number, the authors recommend the use of a stereomicroscope and a magnification ratio of between 64 and 160×. A clear distinction can normally be made between true nodules, mineralised or not, and sites of non-specific mineral deposition. Nodules are typically surrounded by AP+cells and are obviously three-dimensional when compared with other colonies or clusters of cells (Fig. 6.1).

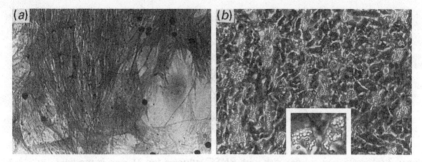

Fig 6.2.(a). Photomicrograph of RMSC cultured for 21 days in medium containing ascorbic acid, β-GP and dexamethasone, stained with von Kossa's reagent, for AP activity, and with a toluidine blue counterstain. An AP positive (dark grey) colony with a fibroblastic, non-bone morphology is shown. An occasional small, toluidine blue positive (black) haematopoietic cell is also seen. Magnification ×300.
(b). Portion of a live fat colony, seen in phase contrast, of RMSC, grown as in (a). The lipid droplets in individual adipocytes are readily apparent at low magnification (×150) and, in the inset, at high magnification (×300).

Identification of non-osteoblastic cells in cultures of RMSC

Haematopoietic cells

Materials

Monoclonal antibodies ED1 and ED2 (Serotec Ltd, Toronto, Canada)
α- naphthyl butyrate esterase kit (Sigma Diagnostics, Sigma, St Louis, MO, US, No. 181–B)

In addition to cells of stromal lineages, the original marrow isolate contains a poorly defined population of haematopoietic precursors. As noted above (p. 89), during the initial stages of culture, these play a vital 'feeder role', but with the continued changes of medium their numbers are rapidly depleted. Adherent cells of the monocyte/macrophage series persist, however, and the authors have shown recently that, when RMSC are maintained in secondary culture under conditions that promote the proliferation and differentiation of osteoprogenitors, there is a concomitant increase in the size of the macrophage population (Herbertson & Aubin, 1995). The possibility that these cells may influence the activities of cells of the osteoblast lineage and that they might mediate some of the actions of exogenous factors on nodule formation in this culture system warrants further consideration. Variations in their number between isolates might also account for the considerable

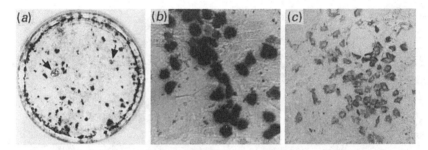

Fig. 6.3.(a). A 35 mm culture dish with RMSC cultured for 21 days in medium containing ascorbic acid, β-GP and dexamethasone, and then stained with α-naphthylbutyrate esterase and a methylene blue counterstain. α-naphthylbutyrate esterase positive (monocyte/macrophage) colonies are visible and the loosely packed labelled cells appear dark grey (arrows). In contrast, tightly packed cells in the bone nodules are methylene blue labelled and appear black. The other colonies, mainly fibroblastic, appear grey in the microphotograph. Magnification (×1.6).
(b). Portion of an α-naphthylbutyrate esterase positive (monocyte/macrophage) colony. Loosely packed cells positive for the enzyme appear black while background fibroblastic cells appear grey. Magnification ×250.
(c). RMSC cultured for 21 days as in (a). Macrophages stained immunohistochemically with anti-rat macrophage antibody ED2 and HRP-labelled second antibody. Cell morphology is similar to that seen in (b), magnification (×150).

variation in the absolute numbers of nodules formed under apparently similar conditions in different experiments.

The presence of cells of the monocyte/macrophage series in cultures of RMSC can be demonstrated (Fig. 6.3), by the use of anti-monocyte/macrophage monoclonal antibodies, ED1 and ED2, which recognise an intracellular and a cell surface epitope, respectively (Fig. 6.3(c)), or alternatively by staining for α-naphthyl butyrate esterase activity (Fig. 6.3(a) and (b)), using kit No. 181-B with the omission of the counterstaining step, which tends to obscure subtle positives.

Adipocytes

Adipocytes are frequently observed in cultures of RMSC, particularly when the medium is supplemented with glucocorticoids, and are also capable of influencing the activities of cells of the osteoblast lineage (Gimble, Chapter 5). The distinctive morphology of these cells makes them easy to identify (Fig. 6.2(b)) and, as a further proof, their content of neutral lipid can be stained using Sudan IV as described below.

Fix the cultures for 1 hour in NBF (p. 99). Rinse once with distilled water and twice (rapidly) with 70% alcohol. Incubate the cultures for 5 to 10 minutes in a solution of Sudan IV (0.1 g w/v prepared in a 1:1 mixture of acetone and absolute alcohol). Discard the Sudan IV solution and differentiate the stain in 70% alcohol for 2–5 seconds. Wash thoroughly in distilled water and allow to air dry. Lipid vacuoles are stained bright red.

Immunohistochemical staining of cultured FRC and RMSC

Materials

Poly-L-lysine 1 mg/ml in H_2O
PBS (p. 96)
10 mM Tris, 150 mM NaCl, pH 7.5 (TBS)
3% H_2O_2 in TBS
3% Bovine serum albumin (BSA)
Horse radish peroxidase (HRP) Colour Development Reagent (Cat. No. 170-6534, Bio-Rad Laboratories Inc., Hercules CA)
Fluorescein (DuPont, Boston, MA)
CY-3 (Jackson Immunoresearch Lab, West Grove, PA)
Moviol (Hoechst Ltd, Montreal, PQ)
UV Epifluorescent microscope (Zeiss Photomicroscope III; Zeiss, Oberkochen, Germany)

Immunohistochemical staining can be performed on cultured FRC and RMSC *in situ* in order to visualise specific intracellular, cell surface and/or extracellular matrix molecules. Using the methods described below, the authors have obtained satisfactory results using a wide variety of antibodies including RBM 211.13, a monclonal anti-rat alkaline phosphatase antibody (Turksen & Aubin, 1991), antibodies recognising components of the extra-cellular bone matrix (osteonectin,osteopontin, osteocalcin, bone sialopro-tein and type I collagen) (Malaval *et al.*, 1994), anti-monocyte/macrophage monoclonal antibodies ED1 and ED2 (pp. 101–102), and an antibody against leukocyte common antigen MRC OX-1 (Herbertson & Aubin, 1995).

The cells can be cultured in standard tissue culture dishes or, if preferred, on glass or plastic coverslips placed in the dishes, or in multi-well slides. To improve the adherence of cells to glass coverslips, they can be pre-washed in acetone or methanol and/or pre-coated with a sterile solution of poly-L-lysine (1 mg/ml in H_2O) for 1–10 minutes at room temperature. At the end of culture, the medium is removed and the cell layers washed once in PBS, prior to fixation for 5 minutes in a solution of 3.7% formalin prepared in

TBS. For intracellular antigens, the cells are then permeabilised by incubation in methanol at −20 °C for 5 minutes.

The following procedure assumes the use of HRP-conjugated secondary antibodies. Block the endogenous peroxidase activity by incubation for 5 minutes at room temperature in a solution of 3% H_2O_2 prepared in TBS. Cover the cells with either the primary antibody or an appropriate control antibody (diluted in TBS containing 3% BSA) and incubate in an humidified atmosphere for 45 minutes at 37 °C. Wash extensively to remove any unbound antibody (3×5 minutes in TBS) and then incubate for a further 30–45 minutes under the same conditions in the presence of the appropriate, HRP-conjugated secondary antibody prepared in TBS. The exact dilution factor will depend on the source of antibody and is usually given by the supplier. Following incubation and removal of the secondary antibody the cells are washed a further 3×5 minutes in TBS. The HRP colour development reagent (4 chloro-1-naphthol, 0.3 % w/v in methanol at −20 °C containing 60 µl of 30% v/v H_2O_2) is mixed immediately before use with an equal volume of TBS and then added to the cultures for 15–60 minutes at room temperature. At the end of this period, the colour development reagent is discarded and the cultures washed in distilled water and allowed to air dry.

The protocol for the immunofluorescent localisation of antigens of interest is essentially the same as that described above except that, after incubation with the primary antibodies, the cultures are incubated in the presence of either fluoroscein- (1/50 dilution) or CY-3- (1/200 final dilution) conjugated secondary antibodies for 30–45 minutes at room temperature. The cultures are then washed to remove unbound antibody, mounted in Moviol and examined by UV epifluorescence microscopy.

Limiting dilution analysis

Materials

Trypsin 0.01% w/v in citrate saline, pH 8.5
96-well microtitre plates (Nunc)

FRC

FRC are isolated as described above (pp. 94–96) and then pre-cultured for 24 hours. At the end of this period, the cultures are washed extensively to

remove debris and dead cells and the viable fraction recovered by trypsinisation (0.01% w/v in citrate saline). The cells are counted, diluted in complete
medium to provide a series of suspensions containing between 25 and 3200
cells/ml and then aliquoted into 96–well microtitre plates at 200 µl/well.

RMSC

RMSC are isolated as described (pp. 96–97) and cultured in complete
medium plus or minus 10 nM dexamethasone, for 7 days prior to analysis.
The cells are harvested by sequential treatment with trypsin and collagenase
and then subcultured into 96–well microtitre plates at between 10 and 750
cells/well.

Analysis of results

The cultures are fixed and stained as described above (pp. 99–100), the
number of mineralised bone nodules present in each well determined with
the aid of a low power microscope and plotted on a semi-log graph.
Assuming that the presence of osteoprogenitors in the samples is rare and
follows a Poisson distribution, their frequency in the original isolate can be
determined by quantifying the fraction of wells not containing nodules at
each of the dilutions tested. Based on transformation of the zero term of a
Poisson distribution: fraction of non–responding wells (ie. wells without
bone nodules)/total wells$= F_0$

$u =$ average number of precursor cells per well
$F_0 = e^{-u}$
$u = -\ln F_0$

Thus the average number of osteoprogenitors per well is equal to the negative natural log of the fraction of wells without a bone nodule.

To achieve acceptable confidence limits, it is essential that a sufficient
number of replicate wells are plated for each dilution of the original cell
suspension. Typically, the authors use a minimum of 96–192 wells (1–2 plates)
for each cell density tested, and present the data as a semi-logarithmic plot of
the fraction of wells not containing bone nodules vs. the number of cells plated.

Influence of other cell types

The limiting dilution assay can be used to assess the influence of other cell
types or their conditioned medium on the proliferation and differentiation

of CFU-O in cultures of RMSC. For this purpose, fibroblasts can be iso-lated from the dermis of 21 day, Wistar rat fetuses by standard explant culture techniques. The cell outgrowths are harvested by trypsinisation at the end of primary or third passage cultures and then subcultured into microtitre plates (10^5 cells/well). The following day, RMSC prepared as described above (pp. 96–97) are then cultured for up to 21 days in the presence or absence of fibroblasts. Alternatively, the non-adherent fraction harvested from cultures of RMSC can be utilised. These are recovered by centrifugation at each change of medium over the first 7 days of primary culture, resuspended in complete medium containing 10% v/v dimethylsulphoxide (DMSO) and then stored frozen at $-70\ °C$. When required, the frozen cells are thawed, washed to remove the DMSO, resuspended in complete medium and then added to microtitre plates (10^5 cells/well) containing RMSC subcultured previously. The cultures are maintained for a total of 21 days. At each change of medium, care should be taken not to disturb the non-adherent fraction. For co-cultures of both fibroblast and non-adherent cells, the fraction of wells not containing nodules at each point on the dilution curve is deter-mined as described above.

Cloning and subcloning

There are several methods that can be used for the generation of clonal cell lines, but the authors have had greatest success using the technique of lim-iting dilution in microtitre plates. To date, this has been applied only to the generation of lines of FRC. The cloning of lines of RMSC by this method, although feasible, is likely to prove more challenging and, for the reasons out-lined above (pp. 88–90), may require the presence of a feeder cell popula-tion or their conditioned medium.

When attempting to clone by limiting dilution, the number of cells added per well should be such that 37% or more of the wells are empty. This is because, based on transformation of the zero term of the Poisson distribu-tion with $F_o = e^{-u}$, and $u = -\ln(0.37) = 0.994$, at the level of 37% non-responding wells, the sample must contain an average of one colony-forming cell (CFC) per well. By the same logic, if 70% of the wells fail to respond (fail to show colony growth), then on average there are only 0.37 CFC cells per sample, 25% of wells with a single CFC and only 5% of wells with two or more cells per well. When generating clones of FRC, the size of the cell inoculum required to satisfy these conditions was estimated from the plating efficiency of cells derived from primary or early passage cultures and, depen-dent on the population, this varied between 0.25 and 10 cells per well. The plates are left undisturbed for between 7 and 14 days and then screened microscopically for evidence of cell growth. Only those wells containing a single colony are selected for further expansion. Thereafter, the cultures are fed every 2–4 days for between 1 and 2 weeks.

Colonies are first subcultured when they comprise about 1000 cells and well before confluence or the formation of a multi-layer. This is done by removing the medium and rinsing the cells with PBS prior to the addition of a few drops of trypsin solution (pp. 94–96). The wells are observed microscopically and, when the cells have begun to round-up, they are recovered by gentle aspiration in a small volume of complete medium and transferred to a larger well. Thereafter, the cells are harvested at, or near, confluence and subcultured at a density of 1×10^4–2×10^4 cells/cm^2. This process is continued until sufficient cells are obtained for experimentation and the generation of frozen stocks (p. 108).

Plating efficiencies of between 5% and 10% are obtained when performing a limiting dilution analysis on FRC from between first and fourth passage cultures. Between 1% and 5% of the cell population have the potential to form colonies of >1000 cells, corresponding to a cumulative population doubling (CPDL) of 9, and 1%–3% have the potential to form colonies >1.5×10^4 cells (CPDL of 12–16). Typically, <1 % of the plated cell population, range 0.1–1.0% dependent on the preparation, are capable of clonal expansion in excess of 10^6 cells (20 CPDL). Of necessity, the above represents only a brief description of our cloning procedure and, for a more detailed consideration, the reader is referred to Aubin et al. (1982).

Replica plating

Materials

Polyester cloth, pore size 1 mm (B&SH Thompson, Scarborough, ON, Canada, Cat. No. HD7–1)

In this approach, a replica or replicas are prepared from colonies on a master dish. The cells remaining on the master dishes are either analysed directly or maintained at a reduced temperature to minimise further proliferation and prevent differentiation. Cells on the replica are transferred to a fresh dish and then cultured at the normal temperature until their developmental potential can be determined. Clones on the master dishes that ultimately gave rise to cells capable of osteogenic differentiation can then be analysed for the expression of known or novel genes utilising a variety of polymerase chain reaction (PCR)-based strategies following the conversion of their mRNA to cDNA using reverse transcriptase (Liu *et al.*, 1994; Candeliere, Liu & Aubin, 1996). For this method, it is essential that the cells are cultured at a density that favours the formation of well-separated colonies. In brief, on day 1 or day 4, a disc of polyester cloth, pore size 1 mm, is placed at the air–liquid interface on the master dish and then submerged and held in contact with the cells by the addition of sufficient 4 mm glass beads to form a monolayer. After a further 4 (day 1) or 7 (day 4) days the polyester 'replica' is recovered and transferred to a new dish. The master dish and the replica are both rinsed carefully with PBS and then fed with complete medium. In experiments where the objective is to prepare mRNA from early osteoprogenitors, the replica disc is cultured at 37 °C, while the master dish is maintained at either 25 °C or 30 °C to arrest the further proliferation and/or differentiation of the cells. Alternatively, both replica filters and master dishes can be grown at the same temperature and analysed concomitantly for progenitor differentiation and responsiveness. Both the replica and master plates receive a change of medium every second or third day.

To determine the transfer efficiency or the fidelity of replication of fetal rat calvaria cell colonies, the replica cloths are fixed on the appropriate day, e.g. day 25 for bone nodule formation, with 10% neutral buffered formalin and stained using the von Kossa technique. Master dishes transferred from the lower temperature to 37 °C or grown continuously at 37 °C are also fixed and stained for comparison. To date, the authors have performed replica plating only on populations of FRC (Liu & Aubin, 1996).

Cryopreservation

Freezing solution is made by mixing 90 ml of α-MEM supplemented with 15% FCS and antibiotics and antimycotics as above (pp. 94–96) with 10 ml dimethylsulphoxide (DMSO). This solution can be stored at −20 °C until required. Cells are frozen in −80 °C at a concentration of 4×10^6 cells per ml of freezing solution. Frozen cells are recovered by rapid thawing in a 37 °C water bath followed by their immediate transfer to a T-75 flask containing 14 ml of pre-warmed culture medium per 1.8 ml of thawed cell suspension ($\sim 7.2 \times 10^6$ cells). The medium containing residual DMSO is removed as soon as the cells have become adherent (typically within 24 hours).

Acknowledgements

Due to limitations of space we were unable to include many 'key' references and instead have cited a number of recent reviews. We acknowledge fully, however, the important contributions made by others to the work discussed in this chapter. The authors studies were supported by grants from the Canadian MRC (MT-12389 and MT-12390). A. H. is in receipt of an MRC Postdoctoral Dental Fellowship.

References

Aubin, J.E. & Liu, F. (1996). The osteoblast lineage. In *Principles of Bone Biology*, ed. J.P. Bilezikian, L.G. Raisz & G.A. Rodan, pp. 51–67. San Diego: Academic Press.

Aubin, J.E. & Turksen, K. (1996). Monoclonal-antibodies as tools for studying the osteoblast lineage. *Microscopy Res. Technique*, **33**, 128–40.

Aubin, J.E., Heersche, J.N.M., Bellows, C.G., Merrilees, M.J. & Sodek, J. (1982). Isolation of bone cell clones with differences in growth hormone responses and extracellular matrix production. *J. Cell Biol.*, **92**, 452–62.

Aubin, J.E., Fung, S-W. & Georgis, W. (1990). The influence of non-osteogenic hemopoietic cells on bone formation by bone marrow stromal populations. *J. Bone Min. Res.*, **5**(2), S81.

Aubin, J.E., Turksen, K. & Heersche, J.N.M. (1993). Osteoblastic cell lineage. In *Cellular and Molecular Biology of Bone*, ed. by M. Noda, pp. 1–45. San Diego & London: Academic Press, Inc.

Aubin, J.E., Liu, F., Malaval, L. & Gupta, A.K. (1995). Osteoblast and chondroblast differentiation. *Bone*, **17**, S77–S83.

Bellows, C.G. & Aubin, J.E. (1989). Determination of the number of osteoprogenitors present in isolated fetal rat calvaria cells *in vitro*. *Dev. Biol.*, **133**, 8–13.

Bellows, C.G., Aubin, J.E., Heersche, J.N.M. & Antosz, M.E. (1986). Mineralised bone nodules formed *in vitro* from enzymatically released rat calvarial cell populations. *Calcif. Tiss. Int.*, **36**, 143–54.

Bellows, C.G., Heersche, J.N.M. & Aubin, J.E. (1990). Determination of the capacity for proliferation and differentiation of osteoprogenitor cells in the presence and absence of dexamethasone. *Dev. Biol.*, **140**, 132–8.

(1992). Inorganic phosphate added exogenously or released from β-glycerophosphate initiates mineralisation of osteoid nodules *in vitro*. *Bone Min*, **17**, 15–29.

Beresford, J.N., Graves, S.E. & Smoothy, C.A. (1993). Formation of mineralized nodules by bone derived cells *in vitro*: a model of bone formation? *Am. J. Med. Genet.*, **45**, 163–78.

Candeliere, G.A., Liu, F. & Aubin, J.E. (1996). Identifying new markers for transitional stages in osteoblast maturation by cDNA fingerprinting of differentiating rat calvaria cell-cultures. *J. Bone Min. Res.*, **11**, M 317.

Falla, N., Van Vlasselaer, P., Bierkens, J., Borremans, B., Schoeters, G. & Van Gorp, U. (1993). Characterization of a 5-fluorouracil-enriched osteoprogenitor population of the murine bone marrow. *Blood*, **82**, 3580–91.

Friedenstein, A.J., Latzinik, N.V., Gorskaya, Y.F., Luria, E.A. & Moskvina, I.L.

110 • Jane E. Aubin and Alexa Herbertson

(1992). Bone marrow colony formation requires stimulation by haematopoietic cells. *Bone and Min.*, **18**, 199–213.

Haynesworth, S.E., Baber, M.A. & Caplan, A.I. (1992). Cell surface antigens on human marrow-derived mesenchymal cells are detected by monoclonal antibodies. *Bone*, **13**, 69–80.

Herbertson, A. & Aubin, J.E. (1995). Dexamethasone alters subpopulation make up of rat bone marrow stromal cell cultures. *J. Bone Min. Res.*, **10**, 285–94.

Lian, J.B. & Stein, G.S. (1995). Development of the osteoblast phenotype: molecular mechanisms mediating osteoblast growth and differentiation. *Iowa Orthop.J.*, **15**(B90), 118–40.

Liu, F., Malaval, L., Gupta, A.K. & Aubin, J.E. (1994). Simultaneous detection of multiple bone-related mRNAs and protein expression during osteoblast differentiation: polymerase chain reaction and immunocytochemical studies at the single cell level. *Dev.Biol.*, **166**, 220–34.

Malaval, L., Modrowski, D., Gupta, A.K. & Aubin, J.E. (1994). Cellular expression of bone-related proteins during *in-vitro* osteogenesis in rat bone-marrow stromal cell-cultures. *J. Cell Physiol.*, **158**, 555–72.

Maniatopoulos, C., Sodek, J. & Melcher, A.H. (1988). Bone formation *in vitro* by stromal cells obtained from bone marrow of young adult rats. *Cell Tissue Res.*, **254**, 317–30.

Turksen, K. & Aubin, J.E. (1991). Positive and negative immunoselection for enrichment of two classes of osteoprogenitor cells. *J. Cell Biol.*, **114**, 373–84.

Van der Plas, A. & Nijweide, P.J. (1992). Isolation and purification of osteocytes. *J. Bone Min.Res.*, **7**, 389–96.

Walsh, S., Dodds, R.A., James, I.E. & Gowen, M. (1994). Monoclonal antibodies with selective reactivity against osteoblasts and osteocytes in human bone. *J. Bone Min.Res.*, **9**, 1687–96.

7

Chondrocyte culture

Ranieri Cancedda, Fiorella Descalzi Cancedda and Beatrice Dozin

Introduction

Cartilage is a dense, specialised connective tissue made of proteins (collagens), polysaccharides and chondrocytes. (For a recent review and selected key references on chondrocyte differentiation see Cancedda, Descalzi Cancedda & Castagnola, 1995). Different types of cartilage exist: hyaline, elastic, and fibrous. In this chapter, only hyaline cartilage will be considered. In the body, hyaline cartilage is found on joint surfaces (articular cartilage), in the few remaining cartilaginous bones (permanent cartilage), and in the cartilaginous models of vertebral column, pelvis, and limb bones that are formed during embryogenesis and subsequently replaced by bone. After birth, this last type of cartilage remains in the growth plate of long bones until sexual maturity is reached. The process by which the cartilaginous model of a bone is at first organised and subsequently replaced by bone tissue, is called endochondral ossification and it is characterised by an orderly sequence of stages of chondrocyte differentiation. Elements of the vertebrate skeleton are initiated as cell condensations. Prechondrogenic mesenchymal cells become closely packed and after establishing cell–cell contacts and gap junctions, start to differentiate into chondrocytes. Thereafter chondrocytes proliferate and secrete increasing amounts of extracellular matrix (ECM) molecules until each single cell is completely surrounded. Soon after the cartilaginous model of the bone is formed, chondrocytes in the central region start to become larger in size (hypertrophic chondrocytes) and to secrete and organise a different ECM. The lowermost part of hypertrophic cartilage undergoes calcification. Deposition of mineral is easily detectable in mammalian growth plates. In avian embryos, the growth plate is poorly organised and the cartilage does not calcify prior to resorption.

Chondrocyte differentiation is regulated by a number of humoral hormones and factors and by locally produced cytokines. Each differentiation stage is characterised by changes in cell proliferation, cell morphology, nature and amount of ECM produced. Collagen is the major component of the ECM. Four of these collagens (II, IX, X, and XI) are considered specific for cartilage. In particular, type X collagen is synthesised by chondrocytes only after they have become hypertrophic and before mineralisation of the ECM occurs. *In vivo*, induction of chondrocyte differentiation and maintenance of the differentiated status strongly depends upon cell–ECM interactions. *In vitro*, cultured chondrocytes change their behaviour and phenotype dramatically when their cell–substratum interactions are modified. There are major similarities in metabolism and gene expression of articular, permanent and growth plate cartilages; however, chondrocytes from these cartilages are cells with a different developmental history and in *in vitro* culture they have different growth requirements. In the literature, data are available on both avian and mammalian cells, although information derived from avian systems cannot always be extrapolated to mammalian systems and vice versa.

Induction and maintenance of the chondrocyte phenotype

Cell–cell interactions and the onset of chondrogenesis

During embryogenesis, onset of chondrogenesis is always preceded by condensation of prechondrogenic cells. A correlation between cell aggregation and formation of cartilage nodules has similarly been observed in culture. In micromass cultures of chick limb bud mesenchymal cells, extensive gap junctions are formed between differentiating chondrocytes, whereas gap junctions are not observed when non-chondrogenic cells differentiate into other connective tissues. The molecular mechanisms, which activate chondrogenesis following cell aggregation, are still unknown.

Cell–substratum interactions and the expression of the chondrocyte phenotype

A decreased cell–substratum interaction strongly affects the behaviour of chondroprogenitors and chondrocytes in culture. Chondrogenesis can be induced easily in prechondrogenic cells by transferring the cells into suspension culture. Spontaneous chondrogenesis occurs in *in vitro* cultures of undifferentiated chick limb cells only when the cells are plated at high cell density or maintained as micromass culture. Similarly, when rabbit or rat growth plate chondrocytes are cultured as a pelleted mass, not only does the chondrocyte phenotype persist, but differentiation also progresses to the hypertrophic chondrocyte stage. Chondrocytes proliferate and are metabolically active in agarose or soft agar. Dedifferentiated rabbit articular chondrocytes and embryonic rat mesenchymal cells express the differentiated phenotype during suspension culture in agarose gel. Chick embryo sternal chondrocytes fully express chondrocyte markers when grown within agar or three-dimensional collagen gels. Articular and growth plate chondrocytes from fetal and adult rabbit, bovine and pig, grow and differentiate when cultured within three-dimensional matrices using calcium alginate beads.

Expression of cartilage specific markers by cultured chondrocytes

Collagens

Changes in the transcriptional activity of various collagen genes and in the steady-state levels of their mRNAs during the *in vitro* differentiation of endochondral chondrocytes have been studied in several culture systems. Type I collagen mRNA is highly expressed both in pre-chondrogenic undifferentiated mesenchymal cells and dedifferentiated chondrocytes, but rapidly decreases during differentiation in culture. The levels of the cartilage specific type II and type IX collagen mRNAs are negligible in the starting cell population, rapidly increase during the first week of culture, reach the maximum level in the second week, and thereafter start to decrease. The expression of the hypertrophic cartilage specific type X collagen mRNA continuously increases until the end of the culture. Differentiation of growth plate chondrocytes from prechondrogenic cells to hypertrophic chondrocytes proceeds through different stages (Castagnola et al., 1986). Stage I (proliferating and differentiating chondrocytes) is characterised by high levels of type II and IX collagens and Stage II (hypertrophic chondrocytes) by very high levels of type X collagen. Stage I may be further subdivided into an early phase (Stage Ia) characterised by a high and transient type VI collagen synthesis and a later phase (Stage Ib) where type II and IX collagens are predominant.

Other proteins

Cultured chondrocytes also express, in a regulated manner, other cartilage specific proteins such as aggrecan, anchorin CII (annexin V), cartilage link protein, cartilage matrix protein, and other proteins expressed by cells of different lineages, which serve as markers of specific chondrocyte differentiation stages. Examples of this last group of proteins are osteopontin, tenascin and the ExFABP (Ch21), a lipocalcin binding long-chain fatty acid, highly expressed by hypertrophic chondrocytes and directly secreted into the culture medium.

Assembly of the extracellular matrix

Ascorbic acid, a co-factor of the prolyl- and lysyl-hydroxylases, is necessary for hydroxylation and assembly of stable collagen fibrils in *in vitro* chondrocyte cultures. Ascorbic acid is an absolute requirement for the correct organisation and maturation of the ECM. The presence of an organised ECM surrounding the cells is required for the expression of certain genes, such as transferrin and retinoic acid–induced heparin binding protein (RIHB).

Matrix mineralisation

In vitro mineralisation of the ECM of cartilage has been observed in cultures of chick embryo tibial chondrocytes, but only when ascorbic acid is added to the culture medium and the ECM is correctly assembled. Mineralisation is preceded by a significant increase in alkaline phosphatase activity. Ultrastructural analysis of the ECM revealed numerous matrix vesicles associated with collagen fibers, considered to be nucleation sites for calcification. In rabbit growth plate chondrocyte culture, calcification of the cartilage formed *in vitro* was induced by hypertrophic chondrocytes themselves and suppressed by TGF-β1 and parathyroid hormone which block chondrocyte maturation.

The influence of soluble mediators on the expression of the chondrocyte phenotype

Peptide growth factors are important in mesoderm induction, in the onset of chondrogenesis during early embryogenesis and in chondrocyte growth and differentiation during endochondral bone ossification. Three major classes of peptide growth factors are of considerable interest. The transforming growth factor β superfamily (which includes the bone morphogenetic proteins or BMPs), the fibroblast growth factor family, and insulin and the insulin-like growth factors. Other soluble mediators known to play an important role in cartilage differentiation include retinoic acid, thyroid hormones, steroid hormones and parathyroid hormone. The effects of many of these agents on chondrocyte proliferation and differentiation in *in vitro* culture is being studied in several laboratories, and major efforts are also being expended on the development of serum-free culture media (Bohme *et al.*, 1992; Quarto *et al.*, 1992). A more detailed consideration of this topic will be found in Cancedda *et al.* (1995).

Phenotypic instability of cultured chondrocytes

In the early stages of differentiation, the chondrocyte phenotype is unstable. When cultured on an adherence permissive substratum, the cells dedifferentiate; they change their morphology and switch from the production of cartilage specific proteoglycans, and cartilage specific collagens, to the production of type I collagen, a characteristic of prechondrogenic cells. Interestingly, these dedifferentiated cells can be expanded in culture and still maintain a chondrogenic potential. Thus, when transferred into permissive culture conditions, dedifferentiated cells will re-acquire the chondrocyte phenotype.

Isolation and culture of chondrocytes

Several cell culture systems have been described which recapitulate the sequence of events observed during differentiation and maturation of chondrocytes *in vivo*. Although there are differences in the starting cell populations and in the culture conditions, in all systems described, a decreased cell–substratum interaction associated with a change in cell shape (rounding-up) is required for induction and/or maintenance of the chondrocytic phenotype. In the following sections, details are given for the procedures currently in use in our laboratory. These include (a) investigation of embryonic growth plate chondrocyte differentiation, and (b) the *in vitro* expansion of articular chondrocytes for use in the repair of damaged joint cartilage.

Primary culture of chick embryonic growth plate chondrocytes

Materials

Phosphate-buffered saline (PBS), pH 7.2 Ca^{2+}- and Mg^{2+}-free
Collagenase II 7 U/ml
Trypsin 0.75 mg/ml
Hyaluronidase 1 mg/ml
Retinoic acid 50–500 ng/ml

Coon's modified Ham F-12 culture medium supplemented with 10% fetal
    calf serum(FCS)
2% chicken serum
1% agarose
40–80 μm nylon mesh
Petri tissue culture dishes

Chick embryo cartilage is commonly used as a source of chondrocytes, as
the cells obtained proliferate and differentiate in culture relatively easily. The
starting tissues used, have included whole tibiae from early embryos (6.5 day;
29–31 H.H. stage), epiphyses of tibiae, caudal and cranial regions of sterna,
and vertebrae, all from late embryos (19 day; 44 H.H.stage)

    The cartilage is carefully dissected free of bone and contaminating soft
tissues under a stereomicroscope. It is then rinsed in PBS and digested with
7U/ml collagenase II and 0.75 mg/ml trypsin in the presence of 2% chicken
serum for 15 min at 37 °C. The supernatant containing tissue debris and
perichondrium is discarded and the cartilage re-digested with the same
mixture (4×40 minutes). Cell dissociation is facilitated by repeated pipetting. The released chondrocytes are pooled, and at this stage are sometimes
filtered through 40–80 μm nylon mesh to remove any residual debris. The
cells are centrifuged at low speed, washed once with PBS, pelleted by
centrifugation and resuspended in culture medium before plating out in
culture (see below).

Pre-chondrogenic to hypertrophic stage

When the freshly dissociated chick embryo chondrocytes from early stage
tibiae are plated in standard tissue culture Petri dishes (adherent culture
conditions), they attach to the plastic and progressively spread assuming a
fibroblastic morphology (Castagnola *et al.*, 1986). At the same time the
chondrocytes stop expressing cartilage specific proteins and switch to the
synthesis of type I collagen and other proteins characteristically expressed by
prechondrogenic cells. When confluent, the dedifferentiated chondrocytes
are trypsinised and passaged, usually twice a week for about 3 weeks. The
culture medium is Coon's modified Ham F-12 supplemented with 10%
FCS.

    When passaged dedifferentiated chondrocytes are transferred into dishes
coated with 1% agarose (non-adherent culture conditions), the cells aggre-
gate rapidly within a few hours and form small clumps. During the first week
in suspension culture, the aggregates increase in size and, when maintained

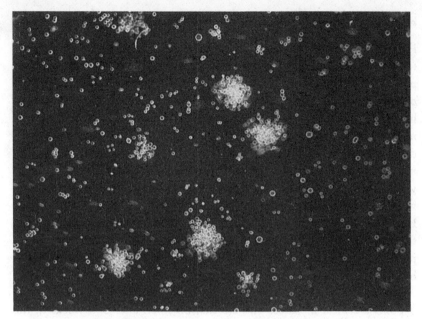

Fig. 7.1. Re-expression of the chondrocyte phenotype by dedifferentiated chick embryo chondrocytes. Primary chondrocytes enzymatically dissociated from 6-day embryo tibiae and dedifferentiated *in vitro* by 3 weeks of adherence culture were transferred to suspension culture in agarose-coated dishes to reinduce the differentiation program. The micrograph illustrates cell maturation after 15 days of suspension culture; cell aggregates formed at the beginning of the culture have started to open, releasing single fully differentiated hypertrophic chondrocytes into the medium.

in the absence of ascorbate, start to release single isolated hypertrophic chondrocytes (Fig. 7.1). After 2–3 weeks, the culture consists entirely of single hypertrophic chondrocytes. Soon after transfer into suspension, the cells synthesise collagen type VI and, shortly afterwards, collagens type II and IX as well as other cartilage specific macromolecules. Later, they also express high levels of collagen type X, which is a specific characteristic of hypertrophic chondrocytes. Standard culture medium does not contain ascorbic acid. When dedifferentiated chondrocytes are cultured as above but in the presence of ascorbic acid, hypertrophic chondrocytes are not released into suspension and small amounts of cartilage are produced (Tacchetti *et al.*, 1987).

Chondrocytes dissociated from late stage tibiae and grown in adherent conditions have a polygonal morphology and a tendency to detach from the dish, giving rise to floating chondrocytes which become hypertrophic.

Hypertrophic chondrocytes to osteoblast-like cells

Hypertrophic chondrocytes are filtered through a 42 μm mesh nylon filter to remove cell aggregates, treated with hyaluronidase (1 mg/ml) and then plated into plastic Petri dishes, which provide an anchorage permissive culture substratum (Descalzi Cancedda *et al.*, 1992). Under these conditions, hypertrophic chondrocytes acquire an elongated or star-shaped morphology, resume cell proliferation, although at a low rate, and they progressively stop depositing an Alcian blue-positive ECM. At the same time, there is an increase in the number of alkaline phosphatase positive cells, especially those with a stellate and/or fibroblastic morphology. Between the first and second week of culture, the cells cease to produce cartilage-specific collagens (types II and X) and initiate the synthesis of collagen type I. Treatment of the cultures between days 1 and 5 with retinoic acid (50–500 ng/ml), which is a potent inducer of cell differentiation, greatly accelerates the transition from hypertrophic chondrocytes to osteoblast-like cells.

Matrix mineralisation in this system occurs post-confluence. The initial deposits are focal in nature and are first observed after 5–6 weeks in the absence of retinoic acid and after 1–2 weeks in its presence. Subsequently, matrix mineralisation spreads throughout the cultures and occurs in association with fibrils of collagen type I. Coincident with these changes the synthesis of osteopontin is detected; a phosphoprotein that has been postulated to act as a bridge between cells and the mineralised matrix. No proteoglycan 'granules' are found in the ECM.

Primary culture of mammalian growth plate chondrocytes

Cultures of mammalian growth plate chondrocytes have been established from cartilage obtained from mouse, rabbit, rat, bovine and human. Cell isolation and culture procedures are basically the same as those described for embryonic chick cartilage, although the enzymatic digestions are often performed for a longer time and, if necessary, repeated. In order to release isolated chondrocytes, some mammalian cartilages also require digestion with hyaluronidase (1–2 mg/ml). In the case of bovine growth plate chondrocytes, the different zones of growth plate cartilage have been cut out and processed separately, in order to obtain chondrocytes at different stages of differentiation (Carey *et al.*, 1993). When rabbit or rat growth plate chondrocytes were cultured as a pelletted mass in minimal medium supplemented with 10% fetal calf serum and 50 mg/ml of ascorbic acid, the cells proliferated for several generations and then reorganised into a cartilage-like tissue that eventually calcified (Kato *et al.*, 1988; Ballock *et al.*, 1993).

Post-natal articular chondrocytes

The development of methods for the isolation and culture of adult chondrocytes has greatly enhanced our understanding of the molecular control of cartilage homeostasis in its fully mature state. It has also permitted the investigation of the mechanism(s) of cartilage degradation that is associated with trauma or chronic disease, e.g. osteoarthritis. In general, however, cultures derived from adult cartilages grow less well and have a shorter life span than those derived from embryonic tissue.

The possibility of producing cartilage-like structures for use in tissue repair and reconstruction has provided a major impetus for the development of adult chondrocyte cell culture models. Rib permanent cartilage and knee joint articular cartilage in particular, have been used as the starting material. In the following paragraphs an attempt has been made to summarise the main guidelines to be followed in order to achieve (a) the successful isolation of articular chondrocytes and (b) their establishment and controlled expansion *in vitro*.

Source of cells

Adult chondrocytes are usually isolated from hyaline cartilage harvested from rib or knee joint articulations. The species mostly used are rabbit, rat and when possible, humans. When obtaining the cartilage sample, the investigator should remember that the biological, biochemical and biomechanical properties of the explant vary dramatically according to its anatomical position. In human adults, for instance, there are major anatomical differences between the joint-bearing surfaces of about 180 synovial joints, all of them being characterised by corresponding variations in microscopic structure and macromolecular composition. Records should be kept of the anatomical location of the explant as well as information concerning age, sex, stature, clinical and dietary history of the donor.

Isolation of cells

Materials

Isotonic saline (NaCl 0.9%)
PBS (p. 115)
Trypsin 0.05–0.25% in 0.02% EDTA
Pronase 1–2 mg/ml
Collagenases I and II 0.2%

Hyaluronidase 1 mg/ml
Ham's F-12 or Dulbecco's modification of MEM (DMEM) supplemented
with 10% FCS
Polytron™ (Kinematica Luzernstrasse 147a, CH-6014, Littau/Luzern
Switzerland)

Preparation of single isolated chondrocytes

Cartilage is extremely sensitive to bacterial degradation and autolysis.
Therefore, the starting cartilage sample must be harvested and handled under
strict aseptic conditions, and processed within 3–5 h of removal from the
body. Moreover, the integrity of hyaline cartilage largely depends upon the
retention in the tissue of normal water content and distribution. For healthy
in vitro culture or metabolic and histological studies, the specimens should
remain fully hydrated. Excised cartilage samples should be rapidly transferred
into a sterile physiological solution (isotonic saline or PBS).

To establish primary chondrocyte cultures, the cartilage must first be dis-
sected with sharp instruments having large, plane cutting surfaces. In this
way, any distortion to which cartilage cells are sensitive can be minimised.
Cartilage samples are first cleaned of any adherent muscular, connective or
subchondral bone tissue, minced into 1–3 mm^3 fragments, rinsed and kept
hydrated in PBS. These fragments are then enzymatically digested in order
to degrade the ECM and to release the chondrocytes. It should be pointed
out that, in mature cartilage, chondrocytes account for only 1% of the total
tissue mass, the major percentage remaining being the collagens and proteoglycans. Accordingly, several alternative protocols have been developed to
digest the matrix, each of them using one or more of the major enzymatic
activities known to act on cartilage matrix. These include:

- non-specific proteases such as trypsin and pronase. Trypsin is normally
  used in the range of 0.05% to 0.25% and can be prepared in 0.02%
  EDTA. Pronase concentration should be 1–2 mg/ml;
- collagen-specific proteases such as collagenases I and II used at concentrations of 0.2%;
- proteoglycan-specific proteases such as hyaluronidase used at 1 mg/ml.

All enzymes should be diluted in PBS and require a constant temperature of
37 °C for optimal activity. For a mild digestion, the cartilage slices should be
extensively rinsed in PBS and transferred into a solution of 0.2% type II collagenase and the suspension gently mixed for up to 16 h using either a
spinner bottle or an orbital shaker stirring at 40–50 rpm. At the end of the

digestion, the cell suspension is gently pipetted and filtered through a sterile nylon membrane (40–150 μm mesh) to eliminate residual undigested fragments. The eluate is centrifuged at 200 g for 10 minutes, and the pellet resuspended in an amount of 0.02% EDTA in PBS equivalent to 40 times the packed cell volume and recentrifuged as above. The final cell pellet is washed twice in PBS and resuspended in an appropriate culture medium (Freed et al., 1995). An alternative, mild digestion protocol requires an initial 30-minute incubation of the cartilage fragments with 2 mg/ml of pronase. This enables the subsequent duration of treatment with collagenase to be reduced and the entire process to be completed within 8 hours (Kirsch, Swoboda & von der Mark 1992).

A rapid digestion protocol that is in routine use in the authors' laboratory involves the use of all three classes of protease mentioned above. Cartilage slices are first subjected to a brief, 20-minute, digestion in PBS containing 0.25% trypsin, 400 U/ml collagenase I, 1000 U/ml collagenase II and 1 mg/ml hyaluronidase. The digested fragments are sedimented by gravity and the released cells in the supernatant discarded, as they may include predominantly non-chondrocytic cells originating from residual bone and soft connective tissue. The pelleted cartilage fragments are then subjected to sequential cycles (1–2 hours) of trypsin/collagenases/hyaluronidase treatment as described above. The fragments and released cells are resuspended every 15 minutes by gentle pipetting. When digestion is complete, the isolated cells are pooled, washed extensively in PBS and then cultured. The standard medium recommended for articular chondrocytes is either Ham's F-12 or DMEM, supplemented with 10% FCS. The choice of culture conditions depends on the particular aspect of cartilage metabolism to be investigated. The most commonly used procedures are summarised on (p. 122).

Preparation of chondrons

In adult hyaline articular cartilage, the chondron consists of a chondrocyte and its pericellular matrix enclosed within a compact, fibrillar capsule. Such chondrons can be extracted and isolated as viable units from low-speed homogenates of fully mature articular cartilage (Poole, Flint & Beaumont, 1988). Chondron culture represents a valuable experimental model for studying the structure, composition and function of the microenvironment that surrounds articular cartilage chondrocytes.

Full-depth tibial cartilage specimens are collected from mature joints. The samples are minced in small fragments, suspended in 20 ml PBS at 4 °C and homogenised for 1–10 min in a Polytrom™ homogeniser (4000–8000 rpm). The homogenate is diluted to 50 ml and subjected to low-speed centrifugation (100g for 30 s) to sediment larger fragments and recover the flocculent supernatant. The sediment is then serially homogenised and centrifuged as above until the sample is exhausted. The pooled supernatants are filtered through a series of nylon filters (decreasing mesh pore size from 1000 to 400 μm) to eliminate unprocessed residual fragments and centrifuged at low speed, 400g for 15–30 minutes. The loose pellet, containing mostly intact chondrons, will remain viable for up to 4 weeks in standard culture medium, either in suspension or in an agarose gel.

Establishment and maintenance of cultures

Primary chondrocytes isolated as described above can be cultured under a variety of conditions; the choice depending on the biological parameters to be analysed. Chondrocytes can be propagated and subcultured as an adherent monolayer or maintained in a three-dimensional environment, such as that provided by culture in suspension over agarose (Fig. 7.2) (Castagnola et al., 1986; Kirsch et al., 1992), within soft agar or agarose (Aydelotte et al., 1986), as a dense cell pellet (Kato et al., 1988; Ballock et al., 1993) or on alginate beads (Ramdi, Legay & Lievremont, 1993). All of these methods are applicable to chondrocytes, embryonal or adult, of any origin or species. In summary, monolayer culture, although allowing rapid and extended proliferation, usually results in the loss of the differentiated chondrocyte phenotype. On the other hand, culture in suspension, by allowing the cells to maintain their spherical shape, favours full expression of the chondrocyte phenotype. It also facilitates major cell–cell interactions and a better integration between the cells and the surrounding ECM they secrete. Thus, monolayer culture is to be preferred if the intention is to obtain an increased number of cells in the shortest amount of time, or if it is to investigate modulation and re-expression of the chondrocyte phenotype. The analysis of the response of chondrocytes to physiological signals and stresses, however, is best investigated in suspension culture.

Potential research applications

The development of articular chondrocyte cultures has recently found a new application, namely the in vitro reconstruction of cartilaginous structures designed for in vivo implantation to treat cartilage defects resulting from pathological situations. In vivo, adult cartilage is characterised by a limited

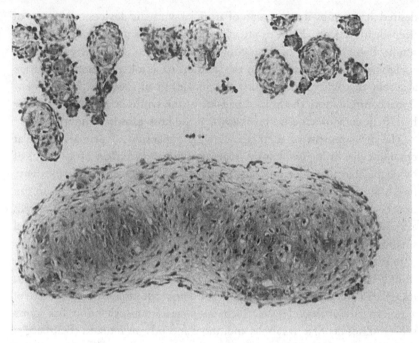

Fig. 7.2. *In vitro* morphogenesis of mammalian cartilage. Human chondrocytes were obtained from knee joint hyaline cartilage by enzymatic digestion of the tissue. After expansion in monolayer culture, the cells were transferred into suspension culture in agarose-coated dishes in medium supplemented daily with 100 µg/ml of ascorbic acid. The differentiating chondrocytes form an organised ECM. After 3 weeks of suspension culture, the cell aggregates were paraffin embedded and 5 µm sections cut and stained with toluidine blue. The micrograph demonstrates the formation of cartilaginous structures characterised by differentiated chondrocytes in lacunae surrounded by a metachromatic ECM.

ability of self-renewal and repair as it is neither vascularised, nor innervated nor penetrated by lymphatic vessels. Consequently, cartilage defects generated from trauma, congenital abnormalities or degenerative joint diseases that are superficial to the subchondral plate do not heal significantly. Only when the lesion reaches across to the vascularised subchondral bone can the cartilage heal to some extent. However, in this case the lesion is usually filled with fibrocartilage that has mechanical properties much lower than those of hyaline cartilage. Sustained damage of the joint eventually leads to osteoarthritis.

Joint failures are currently treated by arthroscopic abrasion and smoothening of the joint area and, when the damage is only superficial, by transplantation of a piece of autologous healthy cartilage after carving it to the desired shape, or by implantation of artificial synthetic devices, in the case of deep-depth defects. These approaches do not always produce the expected results. Difficulties are often encountered, owing to the recurrence of the superficial lesion, limitations in the amount of autologous donor cartilage available, the difficulty in shaping the implant in an adequate three-dimensional configuration, the limited half-life of the synthetic prostheses and the high frequency of adhesive breakdown at the host/prosthesis interface.

The development of methods allowing expansion of primary articular chondrocytes in *in vitro* culture, together with the creation of a variety of biodegradable materials potentially usable as cell delivery systems *in vivo*, have stimulated new therapeutical approaches. Neocartilage formation can now be obtained *in vivo* using either cultures of isolated chondrocytes or cells associated with organic support matrices or with two- or three-dimensional synthetic polymer scaffolds.

In the first approach, primary chondrocytes are harvested arthroscopically from a healthy area of the articular knee joint, serially passaged *in vitro* in monolayer culture and re-implanted as a dense cell suspension at the site of a superficial cartilage lesion. Such autologous transplantation has given successful results first in rabbits and, more recently, in humans (Brittberg *et al.*, 1994). In the second approach, expanded primary chondrocytes are embedded either in fibrin gel or in collagen gel (Wakitani *et al.*, 1989) prior to implantation, thus creating *in vitro* a cartilaginous structure suitable to fill low-depth articular defects. The third approach, although still in its infancy, has virtually unlimited potential, in particular with respect to the repair of large cartilage lesions reaching the subchondral bone. In this system, synthetic biodegradable materials such as polylactic acid (PLA), polyglycolic acid (PGA), Vicryl or Dacron, which can be modulated into a three-dimensional shape to fit the cartilage defect, are seeded *in vitro* with serially passaged chondrocytes. In appropriate culture conditions (Petri dish, spinner flask or bioreactor), the cells proliferate within the scaffold, secrete their own ECM, and a shaped cartilaginous structure is formed ready for implantation. Degradation of the scaffold in parallel with tissue regeneration provides a long-term biocompatible construct. Chondrocyte–polymer constructs of this kind have initially been tested *in vivo* as subcutaneous xenografts in nude mice. Histological analysis of the implant over time has revealed the formation of a structure resembling native hyaline cartilage. More recently, *in vivo* experiments have been successfully performed with cell-PGA implants used as allografts to repair surgically created joint defects in adult rabbit (Freed *et al.*, 1994).

These innovative therapeutical approaches (Fig. 7.3) include several

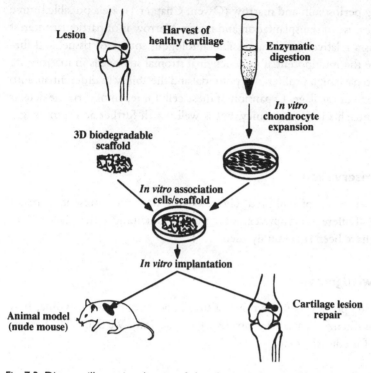

Fig. 7.3. Diagram illustrating the use of chondrocytes for cartilage reconstruction. To repair an injury of joint hyaline cartilage, chondrocytes are harvested from a healthy area of the articular surface and enzymatically dissociated to give a single cell suspension. The chondrocytes are then expanded in *in vitro* adherent cell culture. Following expansion, the chondrocytes are loaded on three-dimensional, shaped synthetic biodegradable polymers and maintained *in vitro* to allow the cells to adhere to, and proliferate within, the scaffold. The chondrogenic potential of the cell/polymer construct is thereafter assessed *in vivo* by subcutaneous implantation in nude mice and further histological analysis of the tissue formed. Constructs with high chondrogenicity can then be implanted *in situ* at the site of the lesion, for cartilage repair.

advantages: (a) a minimal requirement for donor tissue since harvested chondrocytes can be amplified *in vitro*; (b) the possibility, using computerised imaging elaboration, to design the polymeric scaffold, custom-made in terms of three-dimensional configuration and degradation rate; (c) the potential for arthroscopic re-implantation of the cell–polymer constructs which retain their shape and remain malleable.

The multi-potential connective tissue stem cells, which are to be found in tissues of embryonic (Ahrens, Solursh & Reiter, 1977) and adult origin, including periosteum and marrow (Owen, Chapter 1), are a possible source of cells for use in transplantation and repair. Marrow is of particular interest since it is a relatively easily accessible autologous source of tissue, and furthermore the multipotential mesenchymal stromal stem cells in marrow are known to have high proliferative potential and the ability to differentiate into either bone or cartilage. Expansion of these cells for repair of cartilage defects is an approach still in its infancy, but is well worth further active investigation.

Cryopreservation

Cryopreservation of cultured chondrocytes does not present a major problem. Different cryopreservation media containing either glycerol or DMSO have been successfully used.

Acknowledgements

Partially supported by funds from Istituto Superiore di Sanita, Roma, Italy and from Agenzia Spaziale Italiana (ASI), Roma, Italy. We thank Ms. Barbara Minuto for editorial help.

References

Ahrens, P.B., Solursh, M., & Reiter, R.S. (1977). Stage-related capacity for limb chondrogenesis in cell culture. *Dev. Biol.*, **60**, 69–82.

Aydelotte, M.B., Schleyerbach, R., Zeck, B.J. & Kuettner, K.E. (1986). Articular chondrocytes cultured in agarose gel for study of chondrocytic chondrolysis. In *Articular Cartilage Biochemistry*, ed. K.E. Kuettner, R. Schleyerbach & V.C. Hascall, pp. 235–56. New York: Raven Press.

Ballock, T.R., Heydemann, A., Wakefield, L.M., Flanders, K.C., Roberts, A.B. & Sporn, M.B. (1993). TGF beta 1 prevents hypertrophy of epiphyseal chondrocytes: regulation of gene expression for cartilage matrix proteins and metalloproteases. *Dev. Biol.*, **158**, 414–29.

Bohme, K., Conscience-Egli, M., Tschan, T., Winterhalter, K.H. & Bruckner, P. (1992). Induction of proliferation or hypertrophy of chondrocytes in serum-free culture: the role of insulin-like growth factor-I, insulin, or thyroxine. *J. Cell Biol.*, **116**, 1035–42.

Brittberg, M., Lindhal, A., Nilsson, A., Ohlsson, C., Isaksson, O. & Peterson, L. (1994). Treatment of deep cartilage defects in the knee with autologous chondrocyte transplantation. *N. Engl. J. Med.*, **331**, 889–95.

Cancedda, R., Descalzi Cancedda, F. & Castagnola, P. (1995). Chondrocyte differentiation. *Int. Rev. Cytol.*, **159**, 265–358.

Carey, M., Alini, M., Matsui, Y. & Poole, A.R. (1993). Density gradient separation of growth plate chondrocytes. *In Vitro Cell. Dev. Biol.*, **29**, 117–19.

Castagnola, P., Moro, G., Descalzi Cancedda, F. & Cancedda, R. (1986). Type X collagen synthesis during 'in vitro' development of chick embryo tibial chondrocytes. *J. Cell Biol.*, **102**, 2310–17.

Descalzi Cancedda, F., Gentili, C., Manduca, P. & Cancedda, R. (1992). Hypertrophic chondrocytes undergo further differentiation in culture. *J. Cell Biol.*, **117**, 427–35.

Freed, L., Grande, D.A., Lingbin, Z., Emmanual, J., Marquis, J.C. & Langer, R. (1994). Joint resurfacing using allograft chondrocytes and synthetic biodegradable polymer scaffolds. *J. Biomed. Mater. Res.*, **28**, 891–9.

Freed, L.E., Marquis, J.C., Nohria, A., Emmanual, J., Mikos, A.E. & Langer, R. (1995). Neocartilage formation *in vitro* and *in vivo* using cells cultured on synthetic biodegradable polymers. *J. Biomed. Mater. Res.*, **27**, 11–23.

Kato, Y., Iwamoto, M., Koike, T., Suzuki, F. & Takano, Y. (1988). Terminal differentiation and calcification in rabbit chondrocyte cultures grown in centrifuge tubes: regulation by transforming growth factors and serum factors. *Proc. Natl Acad. Sci., USA*, **85**, 9552–6.

Kirsch, T., Swoboda, B. & von der Mark, K. (1992). Ascorbate independent differentiation of human chondrocytes *in vitro*: simultaneous expression of types I and X collagen and matrix mineralization. *Differentiation*, **52**, 89–100.

Poole, C.A., Flint, M.H. & Beaumont, B.W. (1988). Chondrons isolated from canine tibial cartilage: prelimary report on their isolation and structure. *J. Orthop. Res.*, **6**, 408–19.

Quarto, R., Campanile, G., Cancedda, R. & Dozin, B. (1992). Thyroid hormone, insulin and glucocorticoids are sufficient to support chondrocyte differentiation to hypertrophy: a serum-free analysis. *J. Cell Biol.*, **119**, 989–95.

Ramdi, H., Legay, C. & Lievremont, M. (1993). Influence of matricial molecules on growth and differentiation of entrapped chondrocytes. *Exp. Cell Res.*, **207**, 449–54.

Tacchetti, C., Quarto, R., Nitsch, L., Hartmann, D.J. & Cancedda, R. (1987). 'In vitro' morphogenesis of chick embryo hypertrophic cartilage. *J. Cell Biol.*, **105**, 999–1006.

Wakitani, S., Kimura, T., Hirooka, A., Ochi, T., Yoneda, M., Yasui, N., Owaki, H. & Onno, K. (1989). Repair of rabbit articular surfaces with allograft chondrocytes embedded in collagen gel. *J. Bone Joint Surg.*, **71B**, 74–80.

8

Osteogenic potential of vascular pericytes

Ana M. Schor and Ann E. Canfield

Introduction

Pericytes *in vivo*

Pericytes are vascular cells identified *in vivo* by their anatomical location within the subendothelial basement membrane of blood vessels, in particular venules (sometimes called pericytic venules) and capillaries. Such vessels normally consist of three structural elements: (a) the basement membrane, a complex extracellular matrix synthesised by the vascular cells, (b) the endothelial cells, resting on the basement membrane and lining the lumen of the vessel, and (c) the pericytes, periendothelial cells embedded in the basement membrane, except at points of contact with the endothelium. Morphometric studies of skeletal muscle vasculature have indicated that pericytes are randomly positioned in capillaries and concentrated at endothelial cell junctions in venules (Sims, 1991). The ratio of endothelial cells to pericytes is highly variable, depending on species, age, type of vessel, tissue and pathological state (for reviews, see Sims, 1991; Shepro & Morel, 1993).

Pericytes have previously been described as adventitial, mural and Rouget cells. Other cells associated with vessel walls (e.g. smooth muscle cells, reticular cells in the bone marrow, mesangial cells in the kidney, stellate or Ito cells in the liver), cannot always be distinguished from pericytes, since specific markers for the different cell types have not been described. *In vivo*, pericytes are distinguished from endothelial cells mainly by their location, embedded within the basement membrane of vessel walls, and by the presence of von Willebrand Factor and Weibel Palade bodies in endothelial cells. However, this distinction is sometimes blurred, (pp. 129, 130, 134, 136–138).

Origin of pericytes

The mesenchymal derivatives of the neural crest were studied after transplantation of quail embryo neural primordium grafts into chick embryos (Le Lievre & Le Douarin, 1975). The resulting chimeric embryos were examined at different stages of development, and cells derived from the quail graft or the chick donor were identified by species-specific differences in nuclear structure. In all tissues examined, pericytes were of neural crest origin, whereas endothelial cells were of mesodermal origin. Using grafts of limb buds to study endochondral ossification, it was found that endothelial cells in the vessels of such grafts originated from the host tissue, whereas pericytes and osteoblasts originated from the donor (Jotereau & Le Douarin, 1978). These studies clearly indicate that endothelial cells and pericytes are from different lineages. It has, however, been suggested that, in certain processes (e.g. regression of blood vessels during limb morphogenesis in the chick embryo, formation of hypertrophic vessels in the adult) luminal endothelial cells may migrate into peri-endothelial positions, adopting either a pericytic or fibroblastic location and appearance (for review, see Schor, Schor & Arciniegas 1997).

In angiogenesis, new vessels containing endothelial cells and pericytes are formed from the pre-existing vasculature. Whilst it is accepted that the endothelial cells originate from those of the pre-existing vessels, the origin of pericytes is not clear. There are data to suggest that newly formed pericytes may be derived either from those of the pre-existing vessel or from perivascular fibroblasts which migrate towards the new vessels (Schor *et al.,* 1997; Nehls & Drenckhahn, 1993; Schlingemann *et al.*, 1991).

Function of pericytes: vessel homeostasis, contractibility and angiogenesis

As an integral part of the vasculature, pericytes are believed to participate in the maintenance of vessel homeostasis through various functions, which include the synthesis and maintenance of the vascular basement membrane, general metabolism and transport, control of permeability and phagocytosis. Changes in pericyte numbers or functions have been associated with various pathological conditions such as diabetes, neoplasia and hypertension (Shepro & Morel, 1993; Canfield *et al.*, 1990; Schor *et al.*, 1992, 1995). Pericytes express contractile proteins *in vivo* and *in vitro* (Sims, 1991; D'Amore, 1990). Cultured pericytes are able to contract a three-dimensional collagen matrix at the same rate as aortic smooth muscle cells, but at a slower rate than fibroblasts (Schor & Schor, 1986).

The development of new blood vessels displays a clear directionality, with sprouts growing towards the source of the angiogenic stimulus. Both cell migration and proliferation normally take place during sprouting. Pericytes participate in angiogenesis; they respond to angiogenic factors *in vivo* and *in vitro* (D'Amore, 1990; Schor *et al.*, 1997) and are found at the leading end of the growing sprouts (Nehls & Drenckhahn, 1993).

The relationship between pericytes, stromal cells and osteogenic precursors

Undifferentiated or primitive mesenchymal cells present in stromal tissues are able to differentiate into fibroblasts, osteoblasts and smooth muscle as well as other cell types. The identity and location of these multipotential cells is not known, though in several studies it has been suggested that pericytes and/or endothelial cells may fulfil this role (Schor *et al.*, 1990, 1997; Sims, 1991; Shepro & Morel, 1993).

In vivo, evidence from a number of morphological studies support the origin of osteoblasts from either pericytes or endothelial cells (Diaz-Flores *et al.*, 1992). In an experimental model of bone healing, a network of glycogen-rich cells surrounded by basement membrane, and therefore classified as pericytes, were found between sprouting capillaries and osteoblasts (Decker, Bartels & Decker, 1995). Occasionally, Weibel–Palade bodies (a marker of endothelial cells) were observed in the glycogen-rich cells. It was concluded that endothelial cells serve as stem cells and that pericytes are progenitor cells of the osteogenic lineage. Other authors have reached similar conclusions (Schor *et al.*, 1990, 1997). *In vitro*, cloned and uncloned populations of pericytes have been reported to differentiate into bone-forming cells. Pericytes capable of forming a mineralised tissue in culture have been derived from bovine retina and bovine brain (Schor *et al.*, 1990, 1992, 1995; Brighton *et al.*, 1992; Canfield *et al.*, 1996) and from human dental pulp and subcutaneous fat tissues (A.M. Schor, unpublished observation).

Time-lapse videomicroscopy has revealed that pericytes are highly mobile cells and that their migration and proliferation is not contact inhibited. In post-confluent pericyte cultures, the formation of three-dimensional structures resembling nodules is observed (Fig. 8.1(*a*), (*b*)). Unlike the nodules that form in cultures of fetal rat calvarial cells (Aubin & Herbertson, Chapter 6), pericyte nodules do not appear to be clonally derived. Rather, they are formed by detachment and retraction of cells in multi-layered regions of post-confluent cultures. Cells within the nodular structures are metabolically active, express alkaline phosphatase and form abundant extracellular matrix;

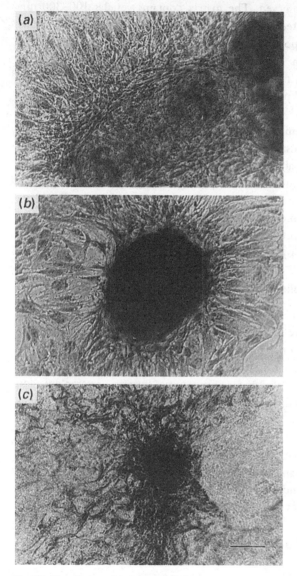

Fig. 8.1. Morphology of bovine retinal pericytes in culture. (*a*) After confluence, the cells first form multilayers and then retract to form nodule-like structures, leaving areas of the substratum cell free or sparsely populated. (*b*) The nodule-like structures subsequently mineralise and appear opaque when viewed under phase contrast microscopy. (*c*) Nodules and multilayers are often connected by long cords of cells; these form most readily when the cells are cultured on a two-dimensional collagenous substratum. (*d*) Pericytes plated within a three-dimensional matrix display an elongated or 'sprouting' morphology. Bar=250 μm for (*a*) and (*b*) and 150 μm for (*c*) and (*d*).

cell necrosis was not observed. The matrix contains vesicles 100–300 nm diameter and cross-striated collagen fibres upon which needle-like crystals of hydroxyapatite are deposited. The presence of mineral can also be demonstrated by staining with Alizarin Red or von Kossa's reagent (Fig. 8.2). Energy-dispersive X-ray microanalysis of this mineral phase has revealed the presence of calcium and phosphorus in a ratio similar to that found in bone (Schor *et al.*, 1990).

Recently, the osteogenic potential of cultured pericytes has been tested in the diffusion chamber assay. The intraperitoneal implantation of chambers containing bovine retinal pericytes in adult athymic mice resulted in the formation of an osteogenic tissue comprising cartilage and bone (Fig. 8.3). In contrast, the implantation of chambers containing bovine fibroblasts resulted in the formation of a soft fibrous tissue, with no evidence of matrix mineralisation (A.E. Canfield, M.J. Doherty, B.A. Ashton & M.E. Grant, unpublished observations).

Potential research applications

The deposition of mineral in soft tissues either as ectopic calcification or bone formation, occurs in many pathological conditions. These include some inherited diseases of the skeleton and conditions where there is a biochemical abnormality resulting in high serum phosphate or calcium or both. Ectopic calcification may also occur after trauma, surgery or paraplegia and in arterial walls in heart disease. The current studies on the origin and function of pericytes, and the conditions under which they promote the formation of a mineralised soft tissue, are relevant to a greater understanding and resolution of these situations. Pericytes are intimately associated with the structure of vessel walls and, as such, are likely to be important in the permeability, contractibility and phagocytic activity of the vascular system and in the pathology of diseases mentioned earlier.

Characterisation of pericytes

In vivo

Pericytes *in vivo* display long filamentous processes that wrap along and around the endothelium; transmission electron microscopy demonstrates the presence of microfilament bundles with focal densities, plasmalemmal vesicles and glycogen deposits. There is no specific marker for pericytes. However, staining with various antibodies, in combination with their location in the vessel, may be used to identify these cells. Pericytes express

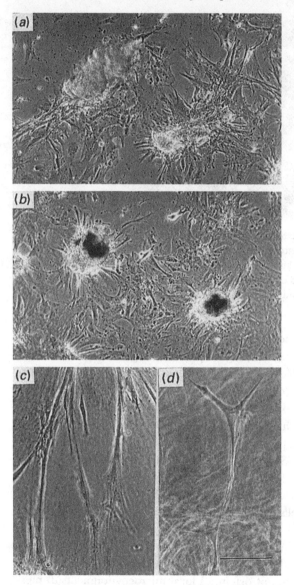

Fig. 8.2. Mineralisation of pericyte nodules *in vitro*. (*a*) Cultures of bovine retinal pericytes contain multilayer areas and (*b*)(*c*) nodule-like structures which show positive staining for alkaline phosphatase (*a*, *b*) and mineral using Von Kossa's reagent (*c*) or Alizarin Red (not shown). Bar=150 μm.

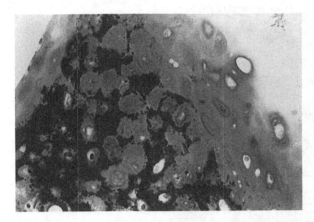

Fig. 8.3. Mineralisation of cultured pericytes implanted *in vivo*. Cultured bovine retinal pericytes were inoculated into diffusion chambers (10^4 to 10^5 cells per chamber) and implanted intraperitoneally into athymic mice under anaesthesia. After 56 days the implants were removed and examined histologically. The presence of mineralised matrix (black) was confirmed by staining with von Kossa's reagent.

desmin, vimentin, α-smooth muscle actin (α-SMA), non-muscle actin, muscle and non-muscle myosin, high molecular weight, melanoma-associated antigen (HMW-MAA), a surface ganglioside antigen (3G5) as well as various proteins (Nayak *et al.*, 1988; Schor *et al.*, 1990; Schlingemann *et al.*, 1991; Sims, 1991; Shepro & Morel, 1993; Nehls & Drenckhahn, 1993).

In vivo, pericytes are heterogeneous with respect to staining with these antibodies. For example, pericytes in pre- and post-capillaries of rat mesentery were positive for α-SMA, while those in mid-capillary were negative (Nehls & Drenckhahn, 1993). In addition, heterogeneity amongst endothelial and smooth muscle cells may complicate pericyte identification. For example, endothelial cells commonly negative for desmin and α-SMA are found to express these proteins in some conditions. The extent and significance of pericyte heterogeneity remains unknown; available evidence suggests it may be related to a variety of factors including: (a) the type of vessel (b) tissue and species of origin (c) local microenvironment (d) presence of cells of different lineage in a pericyte location and (e) the stage of pericyte differentiation (for reviews, see Schor *et al.*, 1990, 1997).

In vitro

Pericytes isolated and grown in an *in vitro* environment display many of the characteristics of their *in vivo* counterparts, including ultrastructural features

Table 8.1. *Phenotypic characteristics of pericytes in culture: a comparison with endothelial cells, smooth muscle cells and fibroblasts*

Characteristic (reference(s))	Cell type			
	Pericyte	Endothelial	Smooth muscle	Fibroblasts
vWF (1–7)	−	+	−	−
GFA (2–4)	−	−	−	−
α-SM-actin (3–6)	+	−	+	±
NM-actin (5,6)	+	+	−	+
SM-myosin (5,6)	+	−	+	−
NM-myosin (5,6)	+	+	−	+
Desmin (2–6)	±	−	±	−
Mab 3G5 (6,7)	+	−	±	−
HMW-MAA (3,4)	+	−	+	−
Response to TGF-β* (3)	−	+	+	±
Gel contraction (1)	+	−	+	+
Gel invasion (1)	−	±	+	+

Notes:
Abbreviations: vWF=von Willebrand Factor; GFA=glial fibrillary acidic protein; SM=smooth muscle; NM=non-muscle; Mab 3G5=ganglioside recognised by monoclonal antibody 3G5; HMW-MAA=high molecular weight, melanoma-associated antigen; +=present; −=not present; ±=conflicting reports in the literature; may be present in certain cells or under certain circumstances.* In semi-confluent cultures. References: (1) Schor & Schor, 1986; (2) Schor *et al.*, 1992; (3) Schor *et al.*, 1995; (4) Schlingemann *et al.*, 1991; (5) D'Amore, 1990; (6) Shepro & Morel, 1993; (7) Nayak *et al.*, 1988.

and antigen expression. These cells cannot be identified by a single specific marker, but are easily distinguished from other vascular or stromal cells by several phenotypic characteristics (Table 8.1). The phenotype expressed by pericytes is modulated by the culture conditions (Schor & Schor, 1986; Schor *et al.*, 1992, 1997; Canfield *et al.*, 1990, 1996). Under standard culture conditions, on two-dimensional substrata, pericytes show a typical stellate morphology with long processes; an example stained with HMW-MAA is shown in Fig. 8.4. These cells are metabolically active and contain abundant glycogen deposits, plasmalemmal vesicles, rough endoplasmic reticulum and mitochondria (Schor *et al.*, 1990).

Cultured pericytes shed debris continuously. At the EM level, this appears to consist of cellular fragments ranging in diameter from 0.3 to 10 μm. Shedding of debris by pericytes is a characteristic of healthy cells (over 95% viable by trypan blue exclusion) and is negligible in senescent cultures. After

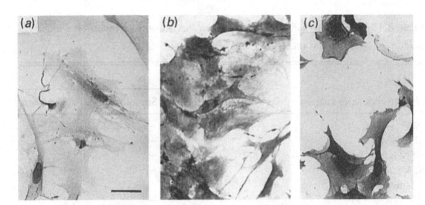

Fig. 8.4. Comparison of human and bovine pericytes: Expression of HMW-MAA. (a)(b) Pericytes isolated from human dental pulp or (c) from bovine retina show morphologic and antigenic similarities. The cells were incubated with either (a) normal mouse serum or a monoclonal antibody recognising the high molecular weight-melanoma associated antigen ((b) and (c)) (HMW-MAA; kindly donated by Dr S Ferrone, New York Medical College). Following extensive washing, the specifically bound primary antibody was detected using a biotinylated secondary antibody, a streptavidin-peroxidase conjugate and diaminobenzidine substrate. Pericytes derived from both sources showed positive staining (compare (b) and (c) with (a)). Bar = 150 μm.

reaching confluence pericytes form nodules, often connected by cellular strands. These nodules can mineralise, appearing opaque when viewed by phase contrast microscopy (Fig. 8.1, (a)–(c)). The development and mineralisation of these nodules is accelerated when the cells are cultured on a native type I collagen gel and the medium is supplemented with 5–10 mM disodium β-glycerophosphate (β-GP; Sigma, G-6251). Conversely, the addition of 1 mM levamisole (Sigma, L-9756), an inhibitor of alkaline phosphatase activity, can inhibit their mineralisation (Schor et al., 1990).

Pericytes cultured within a three-dimensional matrix display an elongated or sprouting phenotype, so called because of its similarity to that of sprouting endothelial cells and the vascular sprouts that form during angiogenesis (Fig. 8.1(d)). Sprouting pericytes and endothelial cells are morphologically indistinguishable and synthesise a similar spectrum of matrix macromolecules (Schor et al., 1992). In spite of these similarities, however, they display distinctive proliferative behaviour when cultured within a three-dimensional macromolecular matrix. Under these conditions sprouting pericytes proliferate whereas endothelial cells do not (Schor & Schor, 1986). The various morphological phenotypes displayed by pericytes in culture are illustrated in Figs. 8.1, 8.2, 8.4 and 8.5(a), (b).

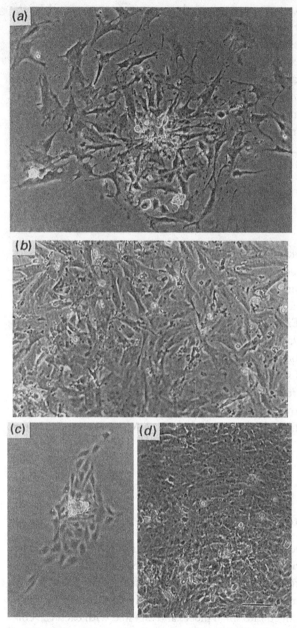

Fig. 8.5. Bovine retinal cells in primary culture. (a) Colony of pericytes growing out of a vessel fragment. (b) Confluent area of pericytes. (c) Colony of endothelial cells. (d) Confluent area of pigmented epithelial cells. Bar = 150 μm.

Cultured pericytes can synthesise both basement membrane and interstitial extracellular matrix proteins, e.g. type IV collagen, laminin, type I and type X-related collagens, tenascin, fibronectin, thrombospondin-1. The biosynthetic phenotype expressed is modulated by the extracellular matrix in contact with the cells and by the stage of cell differentiation, e.g. pericytes growing in monolayer on plastic synthesise collagen types I and III with only traces of type IV, whereas when plated as clumps of cells in gel significant amounts of type IV are secreted into the medium (Canfield *et al.*, 1990, 1996; Schor *et al.*, 1992, 1995).

Isolation and culture of pericytes

Source of cells

Pericytes have been isolated and successfully grown in culture from a variety of different tissues including: retina (D'Amore, 1990; Schor & Schor, 1986; Nayak *et al.*, 1988) brain (Schor *et al.*, 1990), placenta (Challier, Kacemi & Olive, 1995), kidney and liver (Marra *et al.*, 1996), heart (He & Spiro, 1995), adipose tissue (Eskenesy & Tasca, 1988), skin and epididymal fat pad (Shepro & Morel, 1993). The easiest and most readily available source of cells appears to be the bovine retina. Cells isolated from this tissue have been used in the majority of studies reported in the literature. A detailed description of the methods used to isolate pericytes from retinal microvessels will therefore be given below. In addition, a brief description will be given for the isolation of cells from human dental pulp (A.M. Schor, unpublished observations).

Methods of isolation

Bovine retinal pericytes

The procedure described below is used routinely in the authors' laboratories to isolate and characterise retinal pericytes and consists of the following steps. Retinas are removed from bovine eyes and three to six retinas are pooled to make a cell preparation. Four preparations can easily be done at the same time. The retinas are minced into small pieces and digested with bacterial collagenase, trypsin and/or dispase. Microvessels can then be isolated from the digested tissue by filtration through sieves of different pore sizes or by Percoll density gradient centrifugation (D'Amore, 1990; Schor & Schor, 1986; Challier *et al.*, 1995). However, this step is routinely omitted as we have found that vascular cells are normally the only cell type that grow under our culture conditions, irrespective of whether the cell preparation starts with isolated vessels or with the whole tissue digest. The digested tissue is then plated on to Petri dishes and the dishes are incubated at 37 °C. This procedure usually results in the production of a mixed population of cells (pericytes and endothelial cells), which tend to grow from small fragments of microvessels attached to the Petri dish. The yield and relative proportions of pericytes and endothelial cells present in the primary cultures can vary considerably from one cell preparation to another. Several methods can then be used to select pericytes from these initial cultures including ring cloning, the use of specific growth factors and/or culture substrata, weeding, selective trypsinisation and fluorescence-activated cell sorting (FACS) using specific antibodies (D'Amore, 1990; Schor & Schor, 1986; Nayak et al., 1988).

Dissection tools

Spray containing 70% alcohol
Forceps: 1 pair; blunt-ended for collecting retinas
Scissors: 2 pairs with curved blades
Size 21 scalpel blades and holder for dissecting eyes
Dissection board (or plastic sheet)
Six 100 mm diameter Petri dishes (per cell preparation) for washing retinas.
    To five of these, add approximately 10 ml of washing medium.

All equipment is sterilised prior to use. It is useful to have a 100 ml beaker containing 70% alcohol and another beaker with washing medium; these are used to disinfect and wash the dissection forceps and scissors as required.

Collagenase solution
0.5 mg/ml bacterial collagenase type I (Worthington Biochemical Corporation, CLS1) in Minimum Essential Medium with Earle's salts (MEM) (Gibco, 21090–022). The collagenase solution is made fresh as required and filter-sterilised (0.2 μm filter) before use.

Trypsin–EDTA
Provided as a 1× strength solution containing 0.5 g trypsin and 0.2 g EDTA per litre of Modified Puck's Saline A (Gibco, 45300–019). Alternatively, EGTA may be used instead of EDTA (Schor & Schor, 1986).

Washing media
MEM or Hanks balanced salt solution (HBSS; Gibco 24020–091) containing penicillin and streptomycin (200 units/ml and 200 μg/ml, respectively; Gibco, 15140–114).

Growth medium

MEM containing 1 mM sodium pyruvate (Gibco, 11360–039), non-essential amino acids (Gibco, 11140–035), 1 mM glutamine (Gibco, 25030–024), 50 μg/ml ascorbic acid (BDH, 10303) and 20% (v/v) heat inactivated donor calf serum (Gibco, 16030–041) or fetal calf serum (Gibco, 10108–165) (batch tested). The ascorbic acid is added fresh every time the medium is changed; however, a stock solution of 0.5 mg/ml in PBS may be stored in aliquots at −20 °C for up to 3 weeks, with aliquots being thawed only once.

Eye dissection

The dissection should take place in a flow hood under sterile conditions. The dissection board is sprayed with 70% alcohol. The eyes are placed on the board, pupils uppermost, and sprayed generously with 70% alcohol. The back of the eye is held firmly and a coronal incision made behind the vitreous base using a scalpel blade. The eye is cut around using scissors or scalpel. The eye is everted by pushing from the rear with fingers or thumb and the vitreous allowed to slide out. It is often helpful to clamp the diagonally opposed edges of the open eye at this stage. The retina is carefully separated and allowed to hang down with the help of blunt ended forceps and then collected by cutting with scissors, avoiding an area of approx 1 cm circumference around the optic nerve. The retina is placed in a Petri dish containing washing medium. This procedure is repeated with the remaining two to five eyes (placing all the retinas into the same dish). Using the blunt-ended forceps, the retinas are washed by whirling and shaking them vigorously in each of the five Petri dishes containing medium. All traces of choroid should be removed from the retinas by this procedure. Occasionally, it is necessary to remove residual traces of choroid from the retinas using scissors. Finally, the retinas are placed into the empty Petri dish and minced thoroughly using the curved scissors.

Preparation of primary cultures

The procedure for isolation of retinal pericytes is summarised in Fig. 8.6. Freshly prepared collagenase is added to the minced tissue (1 ml/retina) and the 'digestion dish' is placed in a humidified incubator at 37 °C for 4–5 hours. The exact duration of treatment will vary according to the batch of collagenase used and should be optimised for each to obtain the maximum cell yield and viability (Schor & Schor, 1986). The retinas and collagenase are removed from the dish using a 10 ml pipette, placed into a universal, and centrifuged (10 minutes at 150*g*) to pellet the cells and tissue fragments. The supernatant is discarded, and the residue is transferred back to the digestion

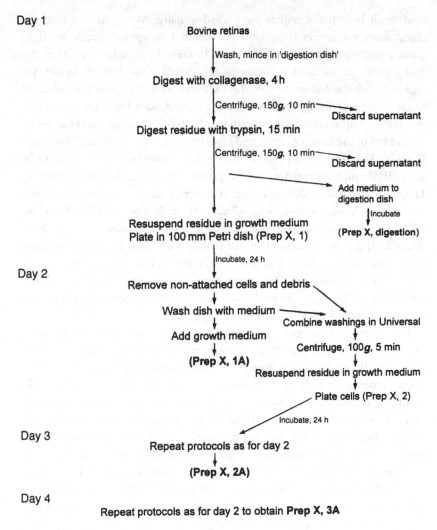

Day 1

Bovine retinas

Wash, mince in 'digestion dish'

Digest with collagenase, 4 h

Centrifuge, 150*g*, 10 min ⟶ Discard supernatant

Digest residue with trypsin, 15 min

Centrifuge, 150*g*, 10 min ⟶ Discard supernatant

Add medium to digestion dish

Incubate

Resuspend residue in growth medium
Plate in 100 mm Petri dish (Prep X, 1)   (Prep X, digestion)

Incubate, 24 h

Day 2

Remove non-attached cells and debris

Wash dish with medium

Add growth medium   Combine washings in Universal

(Prep X, 1A)   Centrifuge, 100*g*, 5 min

Resuspend residue in growth medium

Plate cells (Prep X, 2)

Incubate, 24 h

Day 3

Repeat protocols as for day 2

(Prep X, 2A)

Day 4

Repeat protocols as for day 2 to obtain Prep X, 3A

Fig. 8.6. Flow diagram for the isolation of bovine retinal pericytes.

dish, minced again and incubated with trypsin–EDTA (6 ml) for 15 minutes at 37 °C. The contents of the dish are then transferred into a universal containing 10 ml of growth medium and centrifuged for 10 minutes at 150 *g*. The supernatant is discarded, and the residue is resuspended in approx 6 ml of fresh growth medium and transferred into a Petri dish labelled (PrepX, 1) and incubated overnight. Fresh medium (6 ml) is added to the digestion dish, labelled (**PrepX, digestion**), and this is placed in the incubator. The volume of medium and the size of the Petri dishes used hereafter is dependent upon how many retinas are pooled initially. As a general guide, the digest from six retinas should be resuspended in approximately 6 ml of growth medium and this is plated into a 100 mm Petri dish. The size of the dish should be reduced accordingly if a smaller number of retinas are digested. On the following day, the medium with non-attached cells and tissue is removed from the dish (Prep X, 1), and placed in a universal. The dish is then washed twice with approximately 8 ml of medium and the washings added to the same universal and centrifuged for 5 minutes at 100g. Next (Prep X, 1) is washed carefully a further three times with fresh medium or with HBSS, supplemented with fresh growth medium, relabelled (**Prep X, 1A**), and returned to the incubator. It is important not to wash the dishes too vigorously, otherwise loosely adherent tissue fragments and cells will be dislodged. The cells and tissue fragments recovered by centrifugation from (Prep X, 1) are resuspended in approximately 6 ml of fresh growth medium, transferred to a new Petri dish (Prep X, 2) and then incubated overnight at 37 °C. The next day, the above procedure used for (Prep X, 1) is repeated with (Prep X, 2) to give (**Prep X, 2A**), and the whole process then repeated to obtain (**Prep X, 3A**) (Fig. 8.6).

The purpose of the above procedure is to allow cells and tissue fragments to attach over three consecutive 24-hour periods. It has been found that this maximises the yield of cells from each retinal preparation. Cultures are routinely incubated at 37 °C in an humidified atmosphere consisting of 5% CO_2 in air. All cultures should be monitored daily, and the medium changed if there are substantial amounts of tissue debris present. The cultures normally contain single cells as well as small microvessel fragments out of which cells grow. The amount of cells present in the dishes varies from preparation to preparation, depending partly on the batch of collagenase used. The majority of cells are usually found in dishes **2A** and **3A**, while the **1A** dish may contain a few single cells. Pericytes are also normally obtained from (**Prep X, digestion**) after a period of 7–10 days.

Selection of pericytes from mixed cultures

The isolation procedure described above can produce a mixed population of pericytes and endothelial cells. Pericytes initially appear in primary cultures as large isolated cells or as loose colonies of overlapping, stellate cells with ruffled edges, often growing from a central tissue fragment (Fig. 8.5(a)). When confluent (i.e. covering the substratum) they appear as a disorganised overlapping monolayer (Fig. 8.5(b)). As noted above (pp. 130–132), after confluence the pericytes form multi-layers and nodule-like structures (Fig. 8.1(a), (b)). Endothelial cells are initially present in small, well-defined colonies consisting of 5–100 closely apposed cells that are sometimes attached to tissue debris (Fig. 8.5(c)).

The relative proportions of endothelial cells and pericytes in the primary cultures is rather variable. It is possible, however, to select for pericytes by modifying the culture conditions (D'Amore, 1990; Schor & Schor, 1986). Pericytes grow equally well on different substrata and are therefore routinely cultured on plastic tissue culture dishes. By comparison, the growth of endothelial cells is more dependent on the nature of the substratum. Capillary endothelial cells are also more fastidious than pericytes with regard to their serum requirements and their growth can be inhibited by the addition of transforming growth factor-β (TGF-β, 1 ng/ml) to the growth medium. It is possible therefore, to obtain relatively pure preparations of pericytes by plating the cells on plastic dishes, by the careful choice of serum batch and/or by the addition of TGF-β. A further way of increasing both the plating efficiency and growth of pericytes is by the use of low oxygen tension (Schor & Schor, 1986; Brighton et al., 1992). Occasionally, other cell types are also observed in primary pericyte cultures; the most frequent contaminant being pigmented epithelial cells (Fig. 8.5(d)). Cultures containing such mixed populations of cells should be discarded.

The identification of the pericyte phenotype in culture should be based on as many different criteria as possible. For bovine retinal pericytes we routinely use the following: (a) morphology, (b) lack of Factor-VIII related antigen, (c) lack of glial fibrillary acidic protein, (d) presence of α-SMA, (e) presence of HMW-MAA, (f) immunoreactivity with the monoclonal antibody 3G5, (g) ability to contract three-dimensional collagen gels and (h) migratory behaviour on these gels (Schor et al., 1995; Nayak et al., 1988). Specific techniques for obtaining homogenous preparations of pericytes from pericyte-enriched primary cultures are detailed below.

Cloning

Pericytes can be purified by ring cloning (Schor & Schor, 1986). The location of a small selected colony is marked with a pen on the underside of the plastic dish. Neighbouring cells are scraped off with the tip of a glass Pasteur pipette blown out to form a microneedle. The cultures are observed daily and the neighbouring cells cleared away until the selected colony contains between 150 and 200 cells, at which point it can be ring cloned. The culture is washed with HBSS and sterile glass or perspex rings (4–6 mm internal diameter; coated with sterile silicone grease on the base) are placed on top of the selected colonies and pressed down gently. Trypsin-EDTA is then added to the ring until it is approximately one-third full. When the cells detach, typically after 5 minutes, the suspended cells are transferred into a 35 mm dish (termed 'first clone') containing 1 ml of growth medium and left undisturbed for 5–7 days in an incubator. When cell growth is apparent, the ring cloning procedure can be repeated on this 'first clone' dish, but this time selecting a larger colony containing approximately 300 cells. The cells which proliferate on the second dish are termed the 'second clone'. The small number of cells plated on the 'first clone' dish (i.e. 150–200) and the two-step nature of the cloning procedure increases the probability of obtaining clonal cultures derived from a single parental cell.

Weeding
Unwanted cells (or colonies of cells) can be removed by wiping the dish with either a sterile swab or a sterile pipette tip. The dish is then washed three times with HBSS or PBS, fresh growth medium added and the remaining cells left to proliferate. This procedure should be repeated daily until no contaminating cells are visible.

Selective trypsinisation
This procedure is based on the fact that, before reaching confluence, endothelial cells can be detached from culture dishes relatively easily by a short incubation with trypsin-EDTA, and will then re-attach to culture dishes rapidly. By comparison, pericytes are both slow to detach and to re-attach. It should be noted that this procedure should only be attempted on cultures which are not confluent.

Primary cultures of retinal cells are incubated with approx 3 ml of trypsin-EDTA (per 100 mm diameter Petri dish) for 2–3 minutes. The trypsin solution containing the detached cells (mainly endothelial cells) is then removed and replated onto a dish (dish I) containing fresh growth medium. Fresh trypsin is then added to the original culture dish and the incubation continued for a further 5 minutes. Cells which detach from the dish during this second trypsinisation are collected and added to a second dish containing growth medium, (dish II). Growth medium is also added to the original dish. After 20 minutes, the medium containing unattached cells from dish II is transferred to another Petri dish (dish III) and fresh medium added to dish II. The growth medium in all of the dishes is changed on the following day. By following this procedure, three dishes are obtained from the primary culture; dish I will be enriched in endothelial cells, dish II may contain a mixture of cells, and dish III will be enriched in pericytes. The identity and relative proportions of cells in each dish will depend on the number of endothelial cells and pericytes present in the original dish. Some pericytes will also be left on the original dish that was trypsinised, so it is always worth adding fresh growth medium to this dish and allowing them to proliferate. The timings given are for guidance only and it is important that the progress of cellular detachment and reattachment is monitored microscopically.

Fluorescence-activated cell sorting

Retinal pericytes can also be purified by fluorescence-activated cell sorting using a monoclonal antibody 3G5. This antibody recognises a cell surface ganglioside present on pericytes but not endothelial cells, smooth muscle cells or fibroblasts (Nayak et al., 1988).

Trouble shooting

Lack of cells

The tissue may not have been minced well enough or digested sufficiently. The batch of collagenase should be checked for good activity. It may be necessary for a larger volume of collagenase to be used or the incubation period to be extended. If the problem is still not resolved, different digestion mixtures should be tried. The most effective combination used in the authors' laboratory comprises a mixture of collagenase (0.5–2 mg/ml) and dispase (1–2 mg/ml) in MEM for approx 3 h, followed by treatment with trypsin-EDTA as described above.

Primary cultures are enriched in endothelial cells

Other batches of serum should be tried. TGF-β (1 ng/ml) should be added to the growth medium. Incubating the cultures in 3% O_2 should be tried.

Contamination with other cell types

More care should be taken when removing and washing the retinas. Not too much pressure should be applied with the forceps when removing the retinas. The washing steps are particularly important. In the authors' experience, most problems of contamination are a result of inadequate washing.

Human dental pulp pericytes

When attempting to isolate pericytes from sources rich in connective tissue, it is necessary to isolate the microvessels first. This can be achieved by filtration through different pore-size meshes (Schor & Schor, 1986; D'Amore, 1990) or by Percoll gradient centrifugation (Challier et al., 1995); although both procedures frequently seem to result in contamination with fibroblasts. However, it has been possible recently to isolate homogeneous populations of pericytes from human dental pulp using a modification of a method developed for the isolation of epithelial organoids (Schor *et al.*, 1994; A.M. Schor, A. Nice, S. Pazouki & G. Carmichael, unpublished observations). In this method, the tissue is first finely minced and then digested for up to four consecutive 5–12 hour periods. Each digestion is followed by sedimentation and centrifugation in order to separate single cells from tissue (and vessel) fragments. Cells obtained using this procedure are shown in Fig. 8.4.

Establishment of cultures

Pericytes are maintained routinely on plastic tissue culture dishes in MEM additionally supplemented with 20% (v/v) of either donor calf or fetal calf serum (DCS and FCS, respectively) and 50 μg/ml ascorbic acid. Cultures are incubated at 37 °C in a humidified atmosphere consisting of 5% CO_2 and 95% air. On stock cultures, the medium is changed three times weekly.

Long-term maintenance and subculture

Stock cultures are passaged at a split ratio of 1:2 using trypsin–EDTA. It is important to subculture the cells prior to confluence, as it can be difficult to obtain a single cell suspension once the cells have formed multi-layers and nodule-like structures. Cultures are washed twice with HBSS and incubated with trypsin–EDTA for approximately 5 minutes. The detached cells are plated directly into dishes containing growth medium. The medium is replaced the following day. It should be noted that the plating efficiency of pericytes decreases with increasing passage number. Pericytes have been used between passage 1 and passage 13 for studies on angiogenesis, matrix biosynthesis and matrix mineralisation (Schor & Schor, 1986; Schor *et al.*, 1990, 1992; Canfield *et al.*, 1990, 1996). However, for the majority of studies, pericytes are used at passage 1 or 2.

Cryopreservation

Pericytes are harvested by trypsinisation as described above (p. 146). The cell suspension is placed in a universal containing 8 ml of growth medium and the cells recovered by centrifugation at 100g for 5 minutes. The supernatant is discarded and the cells resuspended at approximately $4 \times 10^5 - 8 \times 10^5$ cells/ml in an ice-cold freezing mixture composed of 30% (v/v) FCS or

DCS and 5% (v/v) dimethyl sulphoxide in MEM. Aliquots (1 ml) are placed in cryovials, frozen according to standard tissue culture protocol and then stored in liquid nitrogen or in a -80 °C freezer until required. To recover frozen cells, the cryovials are incubated at 37 °C and the thawed contents transferred rapidly to a Petri dish containing warm growth medium. The medium is changed on the following day.

In general, it is found that pericytes do not recover well after cryopreservation. It is possible that the optimal conditions for their freezing and long-term storage have yet to be elucidated.

Acknowledgements

The financial support of the Cancer Research Campaign, the Arthritis and Rheumatism Council and the Wellcome Trust is gratefully acknowledged.

References

Brighton, C.T., Lorich, D.G., Kupcha, R., Reilly, T.M., Jones, A. R. & Woodbury, R. A. (1992). The pericyte as a possible osteoblast progenitor cell. *Clin. Orthop. Rel. Res.*, **275**, 287–99.

Canfield, A.E., Allen, T. D., Grant, M.E., Schor, S.L. & Schor, A.M. (1990). Modulation of extracellular matrix biosynthesis by bovine retinal pericytes in vitro: effects of the substratum and cell density. *J. Cell Sci.*, **96**, 159–69.

Canfield, A.E., Sutton, A.B., Hoyland, J.A. & Schor, A.M. (1996). Association of thrombospondin-1 with osteogenic differentiation of retinal pericytes *in vitro*. *J. Cell Sci.*, **109**, 343–53.

Challier, J.C., Kacemi, A. & Olive, G. (1995). Mixed culture of pericytes and endothelial cells from fetal microvessels of the human placenta. *Cell Mol. Biol.*, **41**, 233–41.

D'Amore, P.A. (1990). Culture and study of pericytes. In *Cell Culture Techniques in Heart and Vessel Research*, ed. H.M. Piper, pp. 299–314. Springer-Verlag, Berlin.

Decker, B., Bartels, H. & Decker, S. (1995). Relationships between endothelial cells, pericytes and osteoblasts during bone formation in the sheep femur following implantation of tricalciumphosphate-ceramic. *Anat. Rec.*, **242**, 310–20.

Diaz-Flores, L., Gutierrez, R., Lopez-Alonso, A., Gonzalez, R. & Varela, H. (1992). Pericytes as a supplementary source of osteoblasts in periosteal osteogenesis. *Clin. Orthop. Rel. Res.*, **275**, 280–6.

Eskenesy, M. & Tasca, S.I. (1988). Culture of pericytes isolated from rat adipose tissue microvasculature and characterisation of their prostenoid production. *Cell Biol. Int. Rep.*, **12**, 1055–66.

He, Q. & Spiro, M.J. (1995). Isolation of rat heart endothelial cells and pericytes:

evaluation of their role in the formation of extracellular matrix components. *J. Mol. Cell. Cardiol.*, **27**, 1173–83.

Jotereau, F.V. & Le Douarin, N.M. (1978). The developmental relationship between osteocytes and osteoclasts: a study using the quail-chick nuclear marker in endochondral ossification. *Dev. Biol.*, **63**, 253–65.

Le Lievre, C.S. & Le Douarin, N.M. (1975). Mesenchymal derivatives of the neural crest: analysis of chimaeric quail and chick embryos. *J. Embryol. Exp. Morphol.*, **34**, 125–54.

Marra, F., Bonewald, L.F., Park-Snyder, S., Park, I-S., Woodruff, K.A. & Abboud, H.E. (1996). Characterisation and regulation of the latent transforming growth factor-ß complex secreted by vascular pericytes. *J. Cell. Physiol.*, **166**, 537–46.

Nayak, R.C., Berman, A.B., George, K.L., Eisenbarth, G.S. & King, G.L. (1988). A monoclonal antibody (3G5) -defined ganglioside antigen is expressed on the cell surface of microvascular pericytes. *J. Exp. Med.*, **167**, 1003–15.

Nehls, V. & Drenckhahn, D. (1993). The versatility of microvascular pericytes: from mesenchyme to smooth muscle? *Histochemistry*, **99**, 1–12.

Schlingemann, R.O., Rietveld, F.J.R., Kwaspen, F., van de Kerkhof, P.C.M, de Waal, R.M.W. & Ruiter, D.J. (1991). Differential expression of markers for endothelial cells, pericytes, and basal lamina in the microvasculature of tumors and granulation tissue. *Am. J. Path.*, **138**, 1335–47.

Schor, A. M. & Schor, S.L. (1986). The isolation and culture of endothelial cells and pericytes from the bovine retinal microvasculature: a comparative study with large vessel vascular cells. *Microvacs. Res.*, **32**, 21–38.

Schor, A.M., Allen, T.D., Canfield, A.E., Sloan, P. & Schor, S.L. (1990). Pericytes derived from the retinal microvasculature undergo calcification *in vitro*. *J. Cell Sci.*, **97**, 449–61.

Schor, A. M., Canfield, A.E., Sutton, A.B., Allen, T.D., Sloan, P. & Schor, S.L. (1992). The behaviour of pericytes *in vitro*: relevance to angiogenesis and differentiation. In *Angiogenesis Key Principles – Science – Technology – Medicine*, ed. R. Steiner, P.B. Weisz & R. Langer, pp. 167–78. Basel, Switzerland: Birkhauser Verlag.

Schor, A. M., Rushton, G., Ferguson, J.E., Howell, A., Redford, J. & Schor, S.L. (1994). Phenotypic heterogeneity in breast fibroblasts: functional anomaly in fibroblasts from histologically normal tissue adjacent to carcinoma. *Int. J. Cancer*, **59**, 25–32.

Schor, A. M., Canfield, A.E., Sutton, A.B., Arciniegas, E. & Allen, T.D. (1995). Pericyte differentiation. *Clin. Orthop. Rel. Res.*, **313**, 81–91.

Schor, A. M., Schor, S.L. & Arciniegas, E. (1997). Phenotypic diversity and lineage relationships in vascular endothelial cells. In *Stem Cells*, ed. C.S. Potten, pp. 119–46. London: Academic Press Ltd.

Shepro, D. & Morel, N.M.L. (1993). Pericyte physiology. *FASEB J.*, **7**, 1013–18.

Sims, D.E. (1991). Recent advances in pericyte biology: implications for health and disease. *Can. J. Cardiol.*, **7**, 431–43.

Index

Printed in the United States
By Bookmasters